New Frontiers in Materials Science

Scrivener Publishing
100 Cummings Center, Suite 541J
Beverly, MA 01915-6106

Next-Generation Computing and Communication Engineering

Series Editors: Dr. G. R. Kanagachidambaresan (kanagachidambaresan@gmail.com) and Dr. Kolla Bhanu Prakash (drkbp1981@gmail.com)

Edge computing has become an active research field supporting low processing power, real-time response time, and more resource capacity than IoT and mobile devices. It has also been considered to effectively mitigate loads on data centers, to assist artificial intelligence (AI) services, and to increase 5G services. Edge computing applications along with the IoT field are essential technical directions in order to open the door to new opportunities enabling smart homes, smart hospitals, smart cities, smart vehicles, smart wearables, smart supply chain, e-health, automation, and a variety of other smart environments. However, any developments are made more challenging because the involvement of multi-domain technology creates new problems for researchers. Therefore, in order to help meet the challenge, this book series concentrates on next-generation computing and communication methodologies involving smart and ambient environment design. It is an effective publishing platform for monographs, handbooks, and edited volumes on Industry 4.0 and 5.0, agriculture, smart city development, new computing and communication paradigms. Although the series mainly focuses on design, it also addresses analytics and investigation of industry-related real-time problems.

Publishers at Scrivener
Martin Scrivener (martin@scrivenerpublishing.com)
Phillip Carmical (pcarmical@scrivenerpublishing.com)

New Frontiers in Materials Science

Interdisciplinary Approaches to Innovation and Technologies

Edited by

Kolla Bhanu Prakash
Professor and Associate Dean [R&D] at K.L. Deemed to be University, Guntur, India

S. V. Ranganayakulu
Dean (R&D), Guru Nanak Institutions Technical Campus, Hyderabad, India

and

K. S. Jagannatha Rao
Pro-Chancellor, KL Deemed to be University, Guntur, India

Scrivener Publishing

WILEY

This edition first published 2025 by John Wiley & Sons, Inc., 111 River Street, Hoboken, NJ 07030, USA and Scrivener Publishing LLC, 100 Cummings Center, Suite 541J, Beverly, MA 01915, USA

For more information about Scrivener publications please visit www.scrivenerpublishing.com.

Wiley Global Headquarters
111 River Street, Hoboken, NJ 07030, USA

For details of our global editorial offices, customer services, and more information about Wiley products visit us at www.wiley.com.

The manufacturer's authorized representative according to the EU General Product Safety Regulation is Wiley-VCH GmbH, Boschstr. 12, 69469 Weinheim, Germany, e-mail: Product_Safety@wiley.com.

Library of Congress Cataloging-in-Publication Data

ISBN 978-1-394-31489-8

Front cover image courtesy of Adobe Firefly
Cover design by Russell Richardson

Set in size of 11pt and Minion Pro by Manila Typesetting Company, Makati, Philippines

Printed in the USA

10 9 8 7 6 5 4 3 2 1

Dedicated to My Parents, Sri. Kolla Narayana Rao &
Smt. Kolla Uma Maheswari
and my Wife, Mrs. M. V. Prasanna Lakshmi

Kolla Bhanu Prakash

I dedicate this work to my family members.

S. V. Ranganayakulu

I dedicated this to my wife A. Indira Rao and my beloved son
Dr. Jayanth K Rao for their support

K. S. Jagannatha Rao

Contents

Preface

Materials science has undergone significant evolution in recent years, owing to advances in nanotechnology, biomedicine, energy storage, and smart materials. This book, *New Frontiers in Materials Science: Interdisciplinary Approaches to Innovation and Technologies*, offers a thorough investigation of these cutting-edge sectors, where novel materials and advanced techniques are shaping the future of technology, medicine, and energy solutions. The chapters in this volume take a multidisciplinary approach, providing insights into the synthesis, characterization, and application of advanced materials, thereby laying the groundwork for future scientific advancements.

The first chapter sets the stage by exploring cutting-edge synthesis processes that are central to modern materials science. The discussion then moves to metamaterials, where the manipulation of light and sound waves has the potential to transform technologies ranging from telecommunications to medical imaging. In the healthcare domain, the intersection of materials science and medicine promises revolutionary outcomes, with chapters examining the profound impact of nanoscopic carbon particles and biogenic silver nanoparticles in biomedicine, including their potential for diagnostics, drug delivery, and therapeutic applications.

A primary focus of this volume is energy storage and conversion, which are critical for sustainable future technologies. The book discusses the latest advances in graphene-based supercapacitors, rare earth elements in photonic materials, and other emerging materials that promise more efficient, high-capacity energy sources. As the world shifts toward cleaner energy, these novel materials have the potential to drive major advancements in storage capacity and conversion efficiency.

Beyond energy and healthcare, this book explores smart materials, which hold the key to developing adaptive, intelligent systems that dynamically respond to environmental stimuli. These materials are being integrated into various industries, from electronics to aerospace, and their future developments are examined in depth.

This volume brings together authoritative and pioneering contributions from leading researchers in the field. Each chapter aims to present both foundational knowledge and cutting-edge breakthroughs, making it a valuable resource for scientists, engineers, and students working in materials science and related fields. This book highlights the intersection of materials, energy, healthcare, and technology—an area more critical than ever as we seek to address global challenges related to sustainability and human health.

We believe this book will serve as a valuable resource for understanding recent developments and the future potential of materials science, stimulating further research and innovation in this rapidly evolving discipline. We extend our gratitude to Martin Scrivener and the team at Scrivener Publishing for their support in bringing this volume to light.

Dr. Kolla Bhanu Prakash
Dr. S. V. Ranganayakulu
Dr. K. S. Jagannatha Rao

Acknowledgments

I would like to say thank you to the Almighty and our parents for the endless support, guidance and love through all stages.

I would like to specially thank Sri. Koneru Satyanarayana, President, K. L. Deemed to be University, India for his continuous support and encouragement throughout the preparation of this book.

I would like to express my gratitude towards all who supported, shared, talked things over, read, wrote, offered comments, allowed us to quote remarks and assisted in editing, proofreading and design; through this book journey. We pay our sincere thanks to the open data set providers.

I am grateful to Martin Scrivener and Wiley-Scrivener Publishing team, who showed us the ropes to creating this book. Without that knowledge I would not have ventured into starting this book, which ultimately led to this. Their trusting in us, their guidance and providing the necessary time and resources, gave us the freedom to manage this book.

Last, but definitely not least, I'd like to thank our readers, who gave us their trust and hope this work inspires and guides them.

Dr. Kolla Bhanu Prakash
Dr. S. V. Ranganayakulu
Dr. K. S. Jagannatha Rao

1

Nanomaterial Synthesis and Its Applications

A. Kalaiyarasan*[1], K. Sankar[2], S. Sundaram[1] and K. Gunasekaran[1]

[1]*Department of Mechanical Engineering, Muthayammal Engineering College, Rasipuram, Tamil Nadu, India*
[2]*Department of Civil Engineering, Muthayammal Engineering College, Rasipuram, Tamil Nadu, India*

Abstract

Pyrolysis is a process by which a solid undergoes a heating process to make gas, char, and oil at 400°C. This happens in the absence of oxygen. The disposal of plastics creates environmental problems. Plastic waste can be modified into other products. Carbon nanotubes (CNTs) are produced from paticles sizes ranging from 10 to 50 nanometers. A quartz tube with an inner diameter of 30 mm and a length of 500 mm constitutes the spray pyrolysis apparatus. The nozzle, with an inner diameter of approximately 0.5 mm, was connected to a precursor solution supply. A water bubbler was used to secure the quartz tube's exterior section. Prior to use, the substrate was thoroughly cleaned in acetone utilizing ultrasonication and deionized water, and an argon blower to dry it finely. At the center of the quartz tube was a quartz boat holding the substrate. In an ordinary experiment, the quartz tube was heated to a reaction temperature and flushed with the gas to remove any air. Using an Ar atmosphere, the precursor solution—plastic pyrolytic oil and ferrocrete mixtures—was sprayed into the quartz tube. In the polystyrene pyrolytic oil, ferrocrete was present at a quantity of approximately 25 mg/ml. The homogenous mixture was created by sonicating the solution for 5 minutes, and Ar moved at a speed of 200 sccm/min. The studies were carried out using a reaction period of 45 minutes for each deposition, at 850–1050°C and 1 atmospheric pressure. Following deposition, the furnace was turned off and Ar gas flow was allowed to cool until it reached room temperature. A homogeneous black buildup was seen on the quartz substrate in the reaction zone. Ultimately, the quartz tub's substrate containing CNTs was taken out. The

**Corresponding author*: kalaianbutamil@gmail.com

Kolla Bhanu Prakash, S. V. Ranganayakulu and K. S. Jagannatha Rao (eds.) New Frontiers in Materials Science: Interdisciplinary Approaches to Innovation and Technologies, (1–16) © 2025 Scrivener Publishing LLC

yield and shape of Multiwall Carbon Nanotubes (MWNTs) were synthesized using ferrocene-mixed carbon precursor and polystyrene pyrolytic oil at a flow rate of 20 mL/h was examined. The mature materials for polystyrene-based Multiwall Carbon Nanotubes (PS-MWNTs) were examined using a scanning electron microscope (Hitachi-3000 H). For the Transmission Electron Microscope (TEM), JEOL-JEM-2010F was utilized. The usual tests were used to evaluate the mechanical, thermal, water absorption, and chemical resistance qualities. Effect of three different temperatures (855°C, 952°C, and 1150°C).

Keywords: Nanotubes, nanomaterials, fiber, composite, chemical reaction

1.1 Introduction

The field of nanotechnology has grown significantly during the past century. Additionally, a wide range of academic topics are now directly or indirectly linked to nanotechnology. Nanotechnology is the synthesis, mixing, categorization, and use of materials and devices through nanoscale shape and size changes. The prefix "nano" appears as a keyword in all streams, even those that promote products. The name "nano" comes from the Latin word "nanus" or the Greek word "nanos," meaning "dwarf." It combines the fields of solid-state physics, chemistry, biosciences, and material science. To fully understand a single subject, one must possess an amalgamation of knowledge from the fields of physics, chemistry, material science, solid-state physics, and biosciences. Almost all fields of research and technology are using nanotechnology more and more. The distinction between nanoscience and nanotechnology is that the former uses technology to control matter at the atomic level to create novel nanomaterials with distinctive properties, while the latter provides information about atom configurations and fundamental properties at the nanoscale [1].

Although the general public may not be aware of nanotechnology's continued presence in daily life, nearly all technical fields are keeping an eye on it. Its broad range of use in the fields of electronics, medicine, engineering, environmental science, military, and security is only expanding (Figure 1.1). Even though this technology has seen a great deal of development, more work remains to be done to produce innovative, creative nanomaterials for a variety of applications that will benefit humanity. Despite their size, experience, and expense, researchers are dedicated to their topic and have a strong desire to further their knowledge. Consequently, the disciplines of electronics and medicine are primarily focused on low-cost device downsizing. In the future, nanotechnology will control every aspect

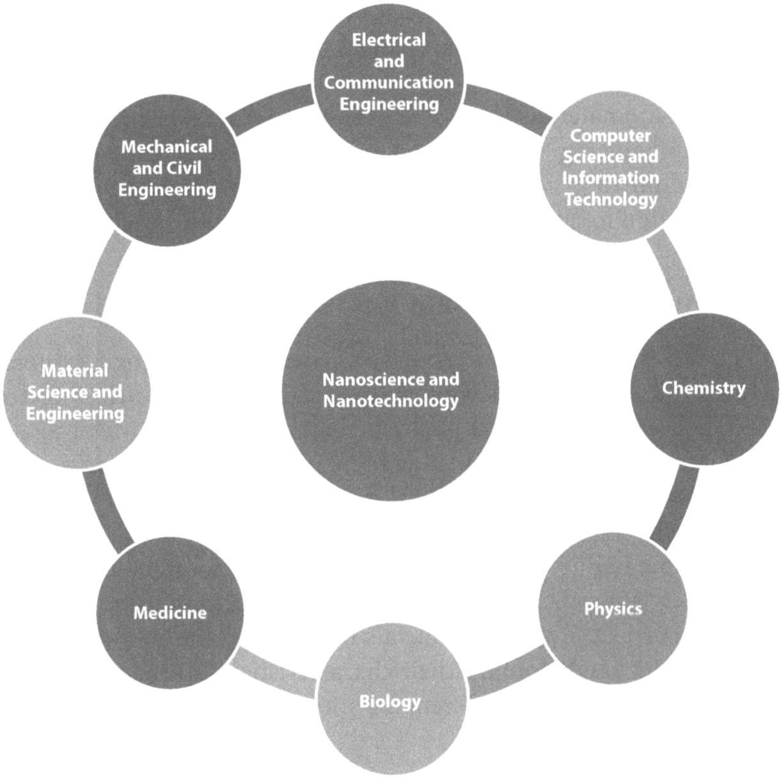

Figure 1.1 Nanotechnology and nanoscience.

of human existence, including working, residing, and communicating. As a result, this generates curiosity about the subject and encourages debate on the fundamental concepts of nanotechnology. "Nanomaterials" are the basic building blocks of nanotechnology. Nanomaterials are defined as materials that are, at least in one dimension, smaller than 100 nm. They are considerably smaller than microscale based on this definition. Usually, nanomaterials are one billionth of a meter, or 10^{-9} m, in size. The fundamental physicochemical characteristics of nanoparticles differ from those of the bulk material depending on their size and form. At the nanoscale level, nanomaterials impressively change in size and shape to acquire new properties and functions. Nanomaterials can be grouped according to their dimensionality and can take on a variety of shapes, including nanorods, nanoparticles, and nanosheets. Nanoparticles are zero-dimensional nanomaterials; nanorods and nanotubes are one-dimensional nanomaterials; and type I films and layers are typically two-dimensional nanomaterials.

Most of them are separated into several categories for individual nanomaterials. Particles undergo physical changes when they come into touch with one another. These fragments of various components are referred to as bulk or 3D nanomaterials. The review highlights the importance of Nano-QSAR in influencing the direction of nano-enabled agriculture going forward, culminating in a synthesis of important observations. It offers tactical direction to guide future research projects in this exciting area.

1.1.1 Nanomaterials Based in Metals

Divalent and trivalent metal ions serve as the foundation for metal nanostructures. Metal nanoparticles may be produced by a variety of techniques, such as chemical and photochemical procedures. Metal ions are reduced to metal nanoparticles using reducing agents. These nanoparticles are very porous and have strong small molecule adsorption capacity. They are extensively employed in many fields of study, including bioimaging and environmental investigations. It is possible to combine two or more nanoparticles while maintaining size control, in addition to achieving a single nanoparticle. Even earth metals can change the properties of the main element through doping. Their characteristics also change when various elements are doped in different constitutions [3–6].

1.1.2 Nanomaterials for Metal and Nonmetals on Semiconductor

Both metallic and non-metallic metal characteristics can be found in semiconductor nanoparticles. By altering them, they display broad band gaps and different characteristics. These are extensively employed in electronic devices and photocatalysis. Examples of semiconductor materials belonging to groups II–VI. Groups III–V include GaN [12], GaP [13], InP [14], and InAs [14]. Recently, researchers have become interested in semiconductor graphene nanocomposites. The semiconductor's chemical and physical characteristics can be enhanced by graphene. The piezoelectric properties of graphene composite materials [18, 19] can be utilized for gas sensing applications [15–17]. Greenhouse gas (GHS) emissions, which are on the rise and have contributed to climate change and global warming [20–23], as well as the depletion of fossil fuel resources, have brought renewable energy sources and energy-saving methods to the attention of scientists and governments worldwide [24–26]. Among the energy sources, transportation accounts for around two-thirds of the world's total energy use [27–29]. For a sustainable future, it is thus imperative to create alternative low-carbon transportation options, such as fuel cell vehicles (FCV) [31], electric vehicles (EV), and hybrid electric vehicles (HEV) [30, 31].

1.1.3 Micro and Nanocomposite Materials

One, two, or three of the phases of the nanocomposite polyphone solid substance have diameters of less than 105 nm. Unlike ordinary composites, nanocomposites have a high surface-to-volume ratio. The following variations in physicochemical qualities may occur depending on size and shape (Table 1.1) [2].

1.2 Nanomaterials and Preparation of Metal Matrix Composite

Nanomaterials are extremely tiny and cannot be observed without a microscope. They range between 1 and 100 nanometers (nm) in diameter. These days, nanoparticles are employed in the production of ceramic coatings, transparent sunscreens, self-cleaning windows, scratch-resistant eyewear, and paints that resist cracking. They are also utilized to make nanocomposites.

Nanoparticles are employed in a variety of sectors, including air purification, environmental preservation, healthcare, and cosmetics. The greatest example of the use of nanoparticles is in the transportation of chemotherapeutic medications. The aircraft industry also uses nanoparticles. There are several advantages of using carbon nanotubes in the morphing

Table 1.1 Actions and ranges of nanoparticles [15].

Range	Action
4 to 6 nm	The catalytic process
<25 nm	Converting the hard magnetic substance to a soft one
<52 nm	To modify the guide of refraction
<100 nm	To attain superparamagnetism
<5nm	The catalytic process
<25 nm	Converting the hard magnetic substance to a soft one

of airplane wings. In a composite form, these kinds of nanotubes are functional. Nanoparticles have a significant role in the sports industry. Parts for Formula One cars can also be made from carbon nanotubes. Manufacturers also use carbon nanotubes in the production of baseball bats. These smaller nanoparticles help athletes perform better since they are lighter. Nanomaterials are also used in the production of mats and towels used by athletes, among other sectors. These goods are all made using antibacterial nanotechnology. They are helpful in preventing bacterially-induced illnesses. The kinds of nanoparticles that have control over their size, shape, and composition are called photovoltaics (PV). Engineers are able to build a variety of solar thermal devices because of these nanoparticles. Every solar product is manufactured to order based on the rates of sun absorption. Nano-titanium dioxide is used by developers to encourage the surfaces of plastic garden chairs to clean themselves. The coating helps to create a sealed layer of water during this procedure. Expansion of the media moving bed is the key to Attritor dry grind processing. We refer to this phenomenon as kinematic porosity. Several forces, including impact, rotation, tumbling, and shear, are applied to the dry particles, making it simple to obtain fine powders in the micron range. The ball milling technique helps to turn the Al_2O_3 into nanopowders, as shown in Figure 1.2. In this experiment, seven 10 mm-diameter tungsten carbide grinding balls are utilized in an Attritor mill. The bowl rotates at 450 revolutions per minute for 1 hour. The first sample has a mesh size of 220. The size of

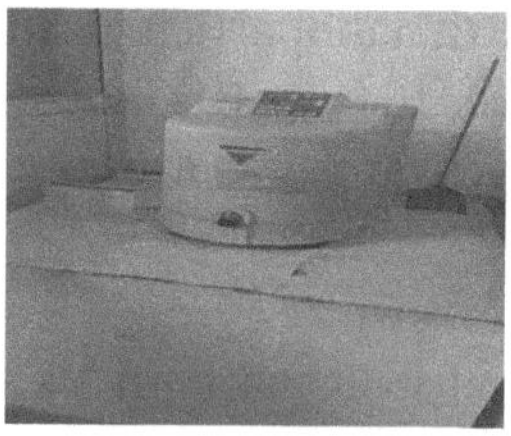

Figure 1.2 Attritor mill.

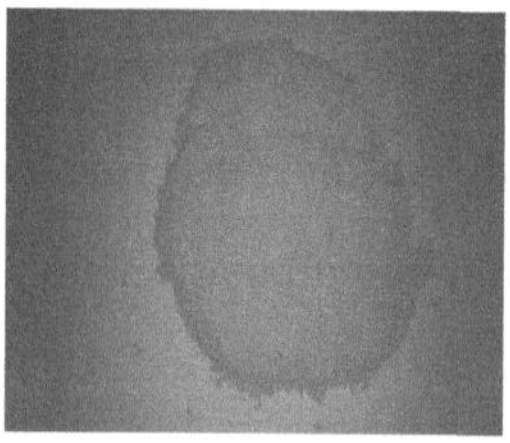

Figure 1.3 Nano WC.

Figure 1.4 Nano TiB_2.

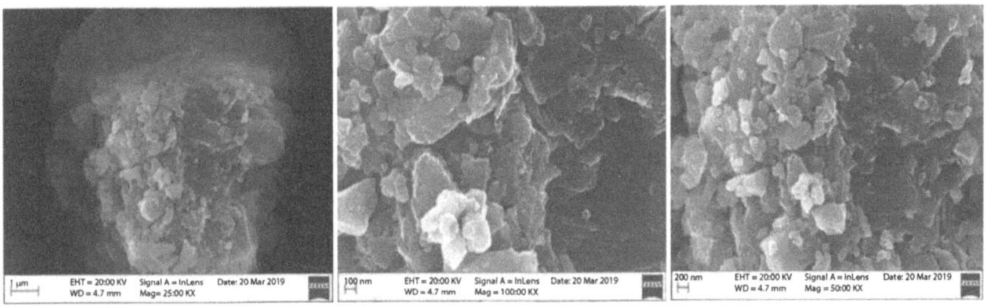

Figure 1.5 FESEM images of nano-Al_2O_3 at 1 μm, 100 nm, and 200 nm scale.

the nanopowder is measured using a FESEM after it has been created. The FESEM images (figure) verify the nanoparticle formation, as depicted in Figure 1.5.

Field Emission Scanning Electron Microscope is abbreviated as FESEM. Instead of using light to see objects, a FESEM uses electrons, which are particles with a negative charge. A field emission source releases these electrons. Electrons scan the item in a zigzag manner. While the electron-generating technology is the primary distinction between SEM and FESEM, FESEM generally operates on the same principles as SEM. A Field Emission Gun (FEG) serves as the electron source in FESEMs. Whereas thermionic emission is employed in SEM, a potential gradient is introduced in FEGs to emit the electron beam.

Alumina nanoparticles are used for making metal matrix composites (MMCs). Stir casting is a process of making an MMC in a liquid state, which is then solidified to produce the MMC. In this process, Al is melted to 720°C and mixed with preheated Al_2O_3, nano-Al_2O_3, and SiC powder. It is properly mixed in molten metal using a stirrer to produce the required MMC, as depicted in Figure 1.6. Cooling is done for nearly half an hour and casting is removed from the mold. A similar procedure is followed for the preparation of SiC composites. One of the main problems that has to be addressed is environmental degradation. Because of their special

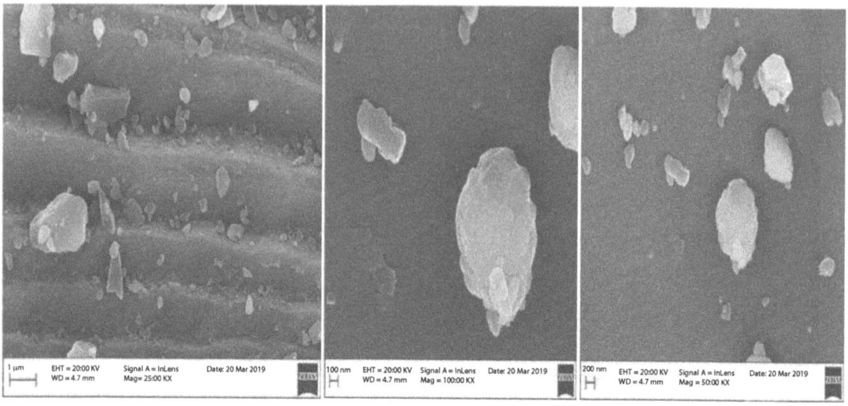

Figure 1.6 FESEM images of nano-SiC at 1 μm, 100 nm, and 200 nm scale.

Figure 1.7 Stir casting.

characteristics, nanomaterials (NMs) have become attractive options for the cleanup of pollution.

1.3 Bio-Medical Applications of Nanomaterials

Nanotechnology has several medicinal applications. The potential for reconstruction and therapeutic advances are quite promising. Today, nanomedicine is advancing. Nanotechnology applications in the medical field are now being implemented across the world. The primary reason

Figure 1.8 Molding.

for this revolution is the scale of the instruments developed to identify and combat sickness, with improved precision and efficiency. This article investigates the medicinal uses of nanoscale materials in the medical field, as well as other important issues. Novel applications in a wide range of industries, including electronics, power, energy, and medicine, have been made possible by the capacity to create nanometric materials with fundamentally different characteristics (optical, electrical, magnetic, thermal, etc.) than macroscopic materials. Nanomedicine, which makes use of the sole property of this material to develop disease identification and action, is the outcome of the application of nanometric or nanostructured materials in medicine. Magnetic nanoparticles are one example of how changes in these materials' fundamental characteristics allow for the development of novel biomedical applications. These materials' distinct qualities, which come from their microscopic size, are complemented by their magnetic nature, making them useful for a range of applications. Among these are: i) altering particle motion to facilitate magnetic separation or drug delivery; ii) producing heat to alleviate hyperthermia in cancer patients; iii) creating gradients of local magnetic fields to enhance contrast in magnetic resonance imaging; and iv) detecting their presence through external methods, opening up applications such as biosensors. Nanomaterials must also fulfill a number of minimal conditions to be used *in vivo*, including being steady at pH = 8, being in a physiological milieu, and not causing an immunological reaction in the body. Because of this, materials that increase the biocompatibility and stability of nanoparticles under physiological settings are often coated on them. Consequently, this coating facilitates the blending of diverse biological materials (medicines, antibodies, etc.) to satisfy the particular needs of each application. It seems reasonable that *in vitro* applications would have less stringent constraints, allowing for the use of other materials or element sizes. Population and construction growth are directly proportional to each other. The concrete manufacturing process cannot be considered an environmentally friendly process.

1.3.1 Advantage of Nanoscale Materials for Biomedical Applications

One of the numerous benefits of using nanomaterials in medicine is their capacity to transfer active substances. Their vast surface area makes it possible to attach therapeutic molecules to them, but they can also be used to encapsulate other substances and improve the ability to protect weakly soluble compounds from degradation processes, change their biodistribution, or increase their solubility. This can be extremely beneficial in combination therapy, enabling the simultaneous delivery of multiple substances across both temporal and spatial dimensions. The ability to more effectively direct compounds toward the area that needs to be treated is another benefit of using nanoformulations. This allows for lower doses of the active ingredient to be used, which in turn lessens the adverse effects of the treatment's systemic distribution, including cancer therapies. Additionally, multifunctional systems can be developed where the disease can be diagnosed and treated with the same particle. Because of all these benefits, the potential biological uses for these materials are increasing quickly, particularly since a single substance may serve several purposes. We have access to nanomaterials for human use in our daily lives; liposomes in cosmetics and titanium dioxide nanoparticles in sun creams are two examples that are widely recognized by the public. Certain nanometric substances, such as iron supplements for the treatment of iron-deficiency anemia (iron oxide nanoparticles are the foundation of Ventofer®, for example), have been used on a daily basis in clinical settings for years. Despite their great potential, only a few nanomaterials have been approved for therapeutic use by drug regulatory organizations. To be more precise, authorized formulations include, among other things, magnetic nanoparticles (Feridex®), albumin nanoparticles (Abraxane®), and liposomes (Doxil®, DaunoXone®). Apart from these, there exist additional highly promising nanomaterials such as quantum dots, polymeric nanoparticles, and gold or silica nanoparticles.

1.3.2 Types of Bio-Nanoparticles Used for the Production of Drugs

The 0.1 or 0.2 micron nanoparticles used in this investigation are composed of biodegradable materials. The systems so obtained are able to sufficiently infiltrate the tissues' cellular and subcellular compartments. To prevent any potential toxicity issues from buildup, we must choose biopolymers that will biodegrade. The two types of nanoparticulate systems that will enable drug molecular targeting are as follows:

1. **Nanocapsules or Reservoir Systems:** They will carry the medication directly to the place of action because they originate from synthetic chemistry.
2. **Matrix Systems:** Synthetic chemistry is where matrix systems originate. The medicine will be distributed in the molecular condition inside this polymer matrix as they entangle in polymer chains. This medium will discharge the medication at the site of action through degradation.

Let's now examine the many uses of nanoscale materials in the medical field:

1.3.3 Biomedical Applications of Nanoscale Materials

The applications of nanoparticles in biomedicine can be broadly classified into three main categories: advancements in nanoparticle-based medicines, sensor-based medicines, and imaging techniques.

1.3.3.1 Bio-Molecular Imaging with Nanoparticle Composite

Research began by providing a succinct summary of the many techniques used to acquire pictures for clinical diagnosis. They also provided some instances of the most popular nanoparticles now in use, such as iron oxide and liposomes. A hybrid material example for nuclear Magnetic Resonance Imaging (MRI) and Positron Emission Tomography/Computerized Tomography (PET/CT) is also included in this section.

1.3.3.2 Nanotherapies in the Field of Biomedical Medicine

To summarize instances of nanoparticle-based treatments. While many different nanomaterials have been the subject of fundamental research for therapeutic applications, only a small number of them have received regulatory agency approval as of yet. However, this will change dramatically in the next years.

1.3.3.3 Optical Nanosensors

Researchers from the University of Zaragoza, including Susana de Marcos, Alba Martín, Jesús Navarro, Isabel Sanz, and Javier Galbán, provided examples of nanoparticles utilized in sensors. These techniques, which are *ex vivo* applications, are not constrained by the toxicity of the nanomaterials, which need careful supervision. With minimal hope for success, research on tissue repair and faculty development is one of the topics receiving the greatest attention. Closer to home, researchers now have additional resources at their disposal thanks to nanotechnology to fight diseases like cancer. Genomic research, bioinformatics, and computer-assisted rational drug design are utilized to create new treatment classes. Currently, research is concentrated on developing ever-more complex materials to optimize their therapeutic use. It wasn't too difficult to create nanoscale materials for nanomedicine until a few years ago. One example would be creating materials whose structure releases active chemicals in response to an external stimulus. This would enable the precise delivery of the target molecule at the appropriate moment to the chosen organ. Internal triggers, such as pH changes or breakdown in the presence of certain tissue enzymes, may cause these materials to become active. They might be triggered by an "external" stimulus such as an oscillating magnetic field or a certain light wavelength. Another example of material development is the functionalization of nanoparticles by the inclusion of different proteins. The challenge here is the vast array of materials and biomolecules that may be mixed together, each having a unique perfect joining method. Finally, studies are being conducted on hybrid materials, which enable the application of many functions at once.

Even though nanomedicine is a broad field of study, there are a number of underlying evolutionary routes that may be identified in its past. On the one hand, it's vital to continue researching the employment of innovative materials, whether they're synthetic or semi-synthetic, as they may provide better or unique features that enable the maximum amount of utilization in a range of applications. However, alternative manufacturing routes must be explored to obtain the materials, whether they are novel or already commercially available. These routes include developing processes that can be profitably and quickly scaled up, as well as developing synthesis methods that produce the least amount of waste. Furthermore, we need to create nanomaterials with the right structural arrangement. It is possible to manufacture several kinds of nanoparticles, such as smart materials and advanced nanomaterials. This overview covers the synthesis, characteristics, and development of several sophisticated nanomaterials as well as recent developments in nanoscience briefly.

When thinking about therapeutic use, it's critical to gain further insight into how biological systems and nanomaterials interact. Understanding how nanoparticles might be able to get past biological barriers is crucial for treating diseases like neurodegenerative or ocular problems, where it might be challenging for drugs to get to the afflicted area. We also need to find out more about the risks nanoparticles pose to human health. Toxicological examinations yield a broad variety of results, and one of the troubles we now face is the lack of standard procedures employed to evaluate the toxicity of nanomaterials during their preclinical evaluation. Any further developments in this area must always be made with the utmost care to ensure the safety of these materials throughout their synthesis, application, and final recycling or disposal. It is critical to lessen the damaging impact of nanoparticles on the environment. One of the challenges is the creation of fully biodegradable materials for biomedical uses that the body can either metabolize or eliminate. Transferring to industry the information gained during the creation of novel nanomaterials is essential for streamlining their production and, in turn, facilitating their clinical use. The medical business has greatly profited from nanoscale materials due to their unique features. Nanoscale materials now have a wide range of applications in medicine; however, further study is required to develop new applications by exploiting existing materials. Examples of these applications include improving techniques for identifying tropical illnesses and creating super adhesives to fix vascular malformations.

1.4 Conclusion

Every day, more novel nanomaterials are produced. Furthermore, synthetic mixed-composition nanomaterials are being developed for various applications. Although they still need to be refined, simple synthesis techniques will provide nanoparticles that are suitable in terms of size, shape, and resistance to environmental changes. There is still room for additional progress in the domains of biomedicine, electronic storage devices, and sensors, despite the amount of research being conducted in these areas at the moment. As a result, this review article will offer a chance to learn more about nanoparticles in general.

Bibliography

1. Shipra Mital Gupta, R.S., *Introduction to Nanotechnology Book*, 1st ed., Oxford University Press India.
2. Vollath, D., *Nanomaterials: An Introduction to Synthesis, Properties and Applications*, 2nd, Publisher, Wiley vch, 2019.
3. Prathap Kumar, M., *et al.*, *J. Environ. Chem. Eng.*, 6, 4, 3907–3917, 2018.
4. Aydın, H., *et al.*, Al-doped ZnO as a multifunctional nanomaterial: Structural, morphological, optical and low-temperature gas sensing properties. Journal of Alloys and Compounds. *J. Alloys Compds.*, 773, 802–811, 2019.
5. Babu, A.T. and Antony, R., Green synthesis of silver doped nano metal oxides of zinc & copper for antibacterial properties, adsorption, catalytic hydrogenation & photodegradation of aromatics. *J. Environ. Chem. Eng.*, 7, 1, 102840, 2019.
6. Faraz, M., Ansari, M.Z., Khare, N., Synthesis of nanostructure manganese doped zinc oxide/polystyrene thin films with excellent stability, transparency and super-hydrophobicity. *Mater. Chem. Phys.*, 211, 137–143, 2018.
7. Reza Rajabi, H., Karimi, F., Kazemdehdashti, H., Kavoshi, L., Fast sonochemically-assisted synthesis of pure and doped zinc sulfide quantum dots and their applicability in organic dye removal from aqueous media. *J. Photochem. Photobiol., B*, 181, 98–105, 2018.
8. Kolahalam, L.A., Viswanath, I.K., Diwakar, B.S., Govindh, B., Reddy, V., Murthy, Y.L.N., Review on nanomaterials: Synthesis and applications. *Mater. Today: Proc.*, 4, 9, 9631–9634, 2017.
9. Kessarin, N. and Ngeontae, W., Circular dichroism glucose biosensor based on chiral cadmium sulfide quantum dots. *Sens. Actuators B Chem.*, 274, 402–411, 2018.
10. Muhammad, F., Tahir, M., Zeb, M., Wahab, F., Kalasad, M.N., Karimov, K.S.,s Cadmium selenide quantum dots: Synthesis, characterization and their humidity and temperature sensing properties with poly-(dioctylfluorene). *Sens. Actuators B Chem.*, 285, 504–512, 2019.
11. Das, S., Banerjee, S., Sinha, T.P., Structural and AC conductivity study of CdTe nanomaterials. *Physica E*, 78, 73–78, 2016.
12. Aydın, C., Variation of structural and electrical properties of gallium nitride-based nanomaterials with nickel dopant at different atomic ratios. *Ceram. Int.*, 44, 14, 17473–17478, 2018.
13. Yue, L., *et al.*, Absorption and fluorescence spectra of gallium phosphide (GaP) nanoparticles. *Trans. Nonferrous Met. Soc. China*, 16, 4, 863–867, 2006.
14. Lee, S.K. and McLaurin, E.J., Recent advances in colloidal indium phosphide quantum dot production. *Curr. Opin. Green Sustain. Chem.*, 12, 76–82, 2018.
15. Wu, H., Li, Q., Bu, X., Liu, W., Cao, G., Li, X., Wang, X., Gas sensing performance of graphene-metal contact after thermal annealing. *Sens. Actuators B Chem.*, 282, 408–416, 2019.

16. Raval, B. and Banerjee, I., *Functionalized graphene nanocomposites and their derivatives*, pp. 295–322, 2019.
17. Sharma, N., Sharma, V., Sharma, S., Sachdev, K., Gas sensing behaviour of green synthesized reduced graphene oxide (rGO) for H2 and NO. *Mater. Lett.*, 236, 444–447, 2019.
18. Li, H., Wang, X., Chen, J., Nonlinear electro-mechanical coupling vibration of corrugated graphene/piezoelectric laminated structures. *Int. J. Mech. Sci.*, 150, 705–714, 2019.
19. Barstugan, R., Barstugan, M., Ozaytekin, I., PBO/graphene added β-PVDF piezoelectric composite nanofiber production. *Compos. B Eng.*, 158, 141–148, 2019.
20. Poorshakoor, E. and Darab, M., Advancements in the development of nanomaterials for lithium-ion batteries: A scientometric review. *J. Energy Storage*, 75, 109638, 2024.
21. Jeffry, L., Ong, M.Y., Nomanbhay, S., Mofijur, M., Mubashir, M., Show, P.L., Greenhouse gases utilization: A review. *Fuel*, 301, 121017, 2021.
22. Galashev, A.Y., Ivanichkina, K.A., Vorob'ev, A.S., Rakhmanova, O.R., Katin, K.P., Maslov, M.M., Improved lithium-ion batteries and their communication with hydrogen power. *Int. J. Hydrogen Energy*, 46, 32, 17019–17036, 2021.
23. Liu, H., Wei, Z., He, W., Zhao, J., Thermal issues about Li-ion batteries and recent progress in battery thermal management systems: A review. *Energy Convers. Manag.*, 150, 304–330, 2017.
24. Warner, J., *The handbook of lithium-ion battery pack design.*
25. Liu, R.J., Yang, L.X., Wang, Y., Liu, H.J., Zhang, X., Zeng, C.L., Effects of carbon nanotubes, graphene and titanium suboxides on electrochemical properties of V2Al1-xCTz ceramic as an anode for lithium-ion batteries. *Ceram. Int.*, 47, 24, 35081–35088, 2021
26. Rao, G. S., and Tsang, J. C., Electrolysis method of intercalation of layered transition metal dichalcogenides. Mater. Res. Bull., 9, 7, 921-926, 1974.
27. Leblanc-Soreau, A., Danot, M., Trichet, L., Rouxel, J., Les intercalaires Ax Ti S2 et Ax Zr S2. Structure et liaisons.(A= Li, Na, K, Rb, Cs). Mater. Res. Bull., 9, 2, 191-197, 1974.
28. Asghar, N., Hussain, A., Nguyen, D.A., *et al.*, Advancement in nanomaterials for environmental pollutants remediation: a systematic review on bibliometrics analysis, material types, synthesis pathways, and related mechanisms. *J. Nanobiotechnol.*, 22, 26, 2024, https://doi.org/10.1186/s12951-023-02151-3.
29. Saleh, T.A., Trends in nanomaterial types, synthesis methods, properties and uses: Toxicity, environmental concerns and economic viability. *Nano-Struct. Nano-Objects*, 37, 101109, 2024.
30. Mohtasham Moein, M., Rahmati, K., Saradar, A., Moon, J., Karakouzian, M., A critical review examining the characteristics of modified concretes with different nanomaterials. *Materials*, 17, 2, 409, 2024.
31. Singh, A.V., Shelar, A., Rai, M., Laux, P., Thakur, M., Dosnkyi, I., Bill, J., Harmonization risks and rewards: nano-QSAR for agricultural nanomaterials. *J. Agric. Food. Chem.*, *72*, 6, 2835–2852, 2024.

2

Advances in Meta-Materials: Engineering Light and Sound Waves for Next-Generation Technologies

N.V. Suresh[1*], Gajalakshmi Sridhar[1], K. Bhanu Prakash[2] and Ananth Selvakumar[1]

[1]ASET College of Science and Technology, Irulapalayam, Kuthambakkam, Chennai, Tamil Nadu, India
[2]CSE Department, KL University, Green Fields, Vaddeswaram, Andhra Pradesh, India

Abstract
Discuss that meta-materials have emerged as a significant advancement in material science, providing fundamental control over light and sound waves. This paper examines recent progress in the design of meta-materials and their implications for new developments at a high level. By precisely controlling the electromagnetic and acoustic properties of these materials, meta-materials enable functionalities that are unimaginable with conventional materials. Key advancements include the development of meta-material–based devices, super-resolution imaging systems, and acoustic central lenses with unprecedented focusing capabilities. Furthermore, meta-materials have changed the fields of optics and acoustics by enabling the development of truly authentic, high-performance devices for applications ranging from broadcast communications to medical imaging. This paper highlights the focal obligations of meta-material evaluation across various fields and explores future opportunities for harnessing these materials in emerging technologies. By figuring out the fundamental principles and recent developments in meta-material science, this paper needs to give an understanding of the remarkable potential for organizing light and sound waves using meta-materials.

**Corresponding author*: sureshnv25@gmail.com

Kolla Bhanu Prakash, S. V. Ranganayakulu and K. S. Jagannatha Rao (eds.) New Frontiers in Materials Science: Interdisciplinary Approaches to Innovation and Technologies, (17–36) © 2025 Scrivener Publishing LLC

***Keywords*:** Meta-materials, light waves, engineering, next-generation technologies, sound waves

Introduction

In the space of materials science and organization, the opportunity for meta-materials has emerged as a moderate perspective, promising superb control over light and sound waves. Meta-materials, deceitfully organized structures with uncommon properties not found in run-of-the-mill materials, have lit gigantic interest and energy in light of their capacity to change various fields, from broadcast communications and imaging to clinical ideas and safeguards. This assessment paper hops into the new advances in meta-materials, examining their essential results to further foster state-of-the-art advances. Meta-materials owe their astounding properties to their versatile microstructures, demandingly coordinated at the nanoscale to show required electromagnetic and acoustic responses (Cerniauskas *et al.*, 2024) [5]. Not a tiny smidgen like normal materials, which get their properties from their compound connection and atomic new development, meta-materials impact organized outlines of sub-wavelength building blocks to achieve surprising functionalities. This granular viewpoint to material strategy enables careful control over limits like refractive record, impedance, and dispersing, expecting exceptional control over the increment of electromagnetic and acoustic waves.

One of the most astounding pieces of meta-materials is their ability to control light in affinities viewed as unbelievable. By exploiting eccentricities like negative refraction, plasmonic resonances, and photonic band gaps, meta-materials can twist, fixate, and steer light with unrivaled precision and capacity (Alù, 2017) [1]. This breaking point holds essential potential for various optical applications, including super-resolution imaging, covering devices, and photonic circuits. Furthermore, meta-material–based optical parts offer novel functionalities far off from standard optics, opening new streets for progress in fields like media communications, perceiving, and quantum developments. Notwithstanding their optical cutoff, meta-materials have comparatively changed the control of sound waves, presenting some other season of acoustical arrangement. By fitting the evaluation and plan of sub-wavelength structures, experts have made meta-materials set up for bowing, demolishing, and arranging sound waves with breath-taking execution (Valipour *et al.*, 2022) [28]. These acoustic meta-materials track down applications across various spaces, from uproar alleviation and sound insurance to ultrasonic imaging and acoustic covering. Moreover, the ability to plan acoustic properties steadfastly enables the creation of moderate and significant acoustic devices with significant functionalities.

The improvements in meta-materials have been driven by interdisciplinary assessment spreading over genuine science, materials science, orchestrating, and math. The obliging energy between speculative encounters, computational appearance, and starter guaranteeing has accelerated the speed of progress, broadening the restrictions of what is reachable with meta-materials. In like manner, joint endeavors between the smart world, industry, and government affiliations have worked with the perception of head openings into obliging applications, driving the commercialization and wide-friendly event of meta-material–based developments.

Looking ahead, the future of meta-materials holds immense commitment and potential. It happened with research endeavors meant to further expand the limitations of meta-materials, pushing toward wonderful functionalities like glorious digestion, overstated dispersing, and dynamic tunability. Additionally, the coordination of meta-materials with emerging advances, such as man-made mental ability, nanofabrication, and 3D printing, opens new edges for improvement and receptiveness. As meta-materials continue to emerge, they are prepared to change various fields, shaping the area of state-of-the-art advancements and opening striking pathways for valid evaluation and mechanical development.

The objective of "Advances in Meta-Materials: Arranging Light and Sound Waves for Cutting-Edge Advances" is to dive into the high-level area of meta-materials to change the control of light and sound waves, thereby driving the improvement of imaginative top-level moves. In particular, the assessment aims to look at the central standards of focal meta-materials, investigating their uncommon properties and approaches to managing acting at the nanoscale level. By understanding these surprising characteristics, the objective is to plan and make meta-material structures custom-fitted for unequivocal applications, utilizing their unprecedented capacity to control the causing of light and sound waves (Imani *et al.*, 2024) [13].

Additionally, the assessment attempts to grow the limitations of organizing by orchestrating novel social event reasoning and materials association methodologies to deal with the effectiveness and adaptability of meta-material creation. This includes investigating new materials with positive properties, further enhancing creation cycles, and organizing apparent-level social gathering activities, such as nanolithography and 3D printing. Besides, the assessment objective consolidates looking at different applications across different fields, including media communications, photonics, acoustics, and clinical imaging. By dealing with the stand-isolated constraints of meta-materials, the point is to help unessential and most mind-boggling execution devices for signal-making due, perceiving, imaging, and communication structures.

In addition, the appraisal attempts to address key inconveniences and endpoints related to current meta-material movements, such as broadband new development, occasion equilibrium, and dynamic tunability. Through an imaginative approach to close and computational appearance, the goal is to overcome these impediments and plan for significant execution in authentic conditions. Also, the evaluation supports interdisciplinary efforts between inspectors, originators, and industry accomplices to speed up the enthusiasm for meta-material examination concerning huge applications. By empowering areas of strength, the goal is to work with information trade, progress move, and commercialization attempts to drive social effect and financial improvement.

Manufacturing limitations keep posing disturbing situations for scaling up meta-materials, from laboratory prototypes to organization programs. High-precision fabrication techniques, which embody 3D printing and nano-imprinting, are being advanced to triumph over those hurdles; however, huge development remains required to make those techniques cost-effective and dependable for large-scale production. Integrating meta-substances into present technology remains complicated because of compatibility issues, cloth stability, and overall performance over time. Addressing those disturbing situations calls for interdisciplinary collaboration at the end of substances science, physics, and engineering to create hybrid answers that merge meta-substances with traditional systems (Zhong *et al.*, 2024) [35].

Energy is no longer an uncommon factor; its overall performance is a crucial area of concern. Many meta-material applications, mainly in optics, require huge electricity inputs to operate optimally. Ongoing studies are focusing on improving passive meta-substances that operate without outdoor electricity, which could notably reduce the electricity footprint of meta-material–based devices. The future of meta-substances lies in the improvement of multifunctional meta-substances, which offer the capability to manipulate both electromagnetic and acoustic waves simultaneously. Exploring the consequences of such substances for applications in communication technology, medical devices, and military stealth technology will provide valuable insights for future advancements.

Literature Review

This study article by Santonocito *et al.* (2023) [21] gives a wide blueprint of steady improvements in meta-materials unequivocally different for optical applications. It covers different centers like strategy methodologies for thinking, creation frameworks, and emerging applications in fields like photonic devices, sensors, and broadcast communications. The article consolidates the huge advances made in controlling light at the nanoscale, including meta-materials, and looks at the likely implications for top-tier optical sorts of progress.

This article by Wu *et al.* (2024) [31] explores the rapidly advancing field of meta-materials applied to acoustic wave control. It delves into theoretical designs for a better understanding of acoustic meta-materials and inspects various procedures for orchestrating acoustic properties, including negative refraction, focusing, and covering. The review also approaches ongoing research efforts and practical applications of acoustic meta-materials in areas such as noise reduction, ultrasound imaging, and cut-down communication.

By focusing on the strong control of meta-material properties, this article dismantles recent advancements in tunable meta-materials and their various applications. It discusses various instruments for achieving tunability, such as outside lifts, stage progresses, and reconfigurable plans. The study highlights key kinds of progress in areas like tunable optics, flexible cover, and reconfigurable receiving wires, showing the constraint of tunable meta-materials to change state-of-the-art developments (Kumar and Mitra, 2023) [14].

This article by Govindaraman *et al.* (2021) [12] investigates the utilization of meta-materials for harvesting and converting energy across different eminent augmentations. It outlines diligent advances in meta-material–based devices for solar energy procuring, thermal management, and electromagnetic energy conversion. The review discusses novel contemplations, such as plasmonic meta-materials for redesigned light trapping and phononic meta-materials for effective thermal management, addressing the restriction of meta-materials to address key challenges in sustainable energy development.

By focusing on the assembly point of meta-materials and perceiving advances, this article gives a framework for late enhancements in meta-material–refreshed sensors. Wu *et al.* (2022) [32] discuss how meta-materials provide exceptional control over electromagnetic and acoustic waves, leading to improved capabilities in areas such as biomedical diagnostics, regular checking, and security applications. The study also examines emerging models in integrated meta-material sensor stages, making plans for moderate, fragile, and multifunctional perceiving devices.

This major overview covers the foundational standards, innovative development, and recent advancements in meta-materials, emphasizing their extraordinary properties for controlling light and sound waves (Askari *et al.*, 2020) [2]. It gives a cautious game plan of meta-material structures, creation strategies, and applications across different fields. The review recalls recent breakthroughs in meta-material–based devices designed for nanoscale light manipulation. It explores advancements in plasmonics, photonic crystals, and meta-surfaces alongside emerging applications in optical communications, sensing, and imaging.

By focusing on the field of acoustic meta-materials, this study explores speculative thoughts, plan strategy, and exploratory presentations for sorting out sound waves. It highlights ongoing updates in acoustic covering, waveguiding, and vibration control, with potential applications in noise reduction, ultrasound imaging, and cut-down communication. This survey examines meta-material–based safeguards expected to capture and convert electromagnetic energy across a wide range of frequencies. It analyzes different maintenance aspects, such as impedance matching, reverberation, and multifaceted structures, along with practical applications in secret development, energy harvesting, and electromagnetic shielding (Liao *et al.*, 2021) [18].

By focusing on the tunability of meta-material properties, this study examines a system for effectively controlling light and sound waves through external enhancements. It covers both dynamic and passive tuning structures, including mechanical, electrical, and thermal change, and analyzes applications in reconfigurable optics, flexible antennas, and programmable acoustic devices. This survey bridges the gap between theoretical models and initial observations of meta-material–based nanophotonic devices (Simovski and Tretyakov, 2020) [22]. It addresses the challenges in design, evaluation system, and considerations necessary to accomplish the desired optical functionalities at the nanoscale, with applications in advanced photonics, quantum optics, and on-chip sensing.

Dal Poggetto (2023) [8] focuses on wave front organizing using metamaterials, examining recent advancements in the manipulation of light and sound fields with noteworthy accuracy. It covers phase change, polarization control, and distortion adjustment structures, addressing inconveniences related to adaptability, data transmission, and creation yield. This study analyzes biomimetic ways of managing the directing meta-material approach, mixed with the amazing plans tracked down in nature. It looks at how ordinary subjects, for example, photonic pearls in butterfly wings and echolocation in bat ears, have mixed novel meta-material structures with redesigned functionalities for sensing, imaging, and cover.

Lvov *et al.* (2022) [19] explore the relationship between meta-materials and quantum impels, analyzing recent endeavors to oversee quantum impacts for controlling light and sound at the nanoscale. It covers quantum meta-materials, quantum plasmonics, and quantum acoustics, with applications in quantum managing, cryptography, and metrology. This survey also highlights the development of meta-material research from the lab to demanded applications, including key irritates and pathways for commercialization. It considers flexibility, cost-effectiveness, and legitimate assessments, as well as making business districts in broadcast communications, healthcare, and renewable energy.

Boardman *et al.* (2011) [4] revolve around late movement in the improvement of dynamic and tunable meta-materials, which can particularly control their optical properties. This overview looks at varied methodologies for achieving tunability in meta-materials and their typical applications in reconfigurable optics. Uriri *et al.* (2020) [26] explore the emerging field of meta-materials for quantum information progress, assessing applications for quantum computing, communication, and sensing. Their framework joins the astounding properties of meta-materials that make them suitable for controlling quantum states of light and matter.

Zhang *et al.* (2023) [34] give a graph of pushing kinds of progress in meta-materials for acoustics, incorporating sound managing, control, and focusing. Their diagram covers a myriad of purposes, from sound reduction to clinical ultrasound imaging. Wang *et al.* (2020) [29] look at the usage of meta-materials for thermal management applications, including heat security, radiation control, and thermal shielding. Their review consolidates the astounding thermal properties of meta-materials and their conceivable impact on energy cutoff and sustainability.

Suresh Kumar *et al.* (2021) [23] review the usage of meta-materials for sensing applications, including biosensing, organized locale, and standard checking. Their review examines the philosophy guidelines and creation methods for meta-material–based sensors. Barri *et al.* (2021) [3] discuss the prerequisite of meta-materials for energy harvesting and management applications, including solar cells, thermoelectric generators, and super capacitors. Their survey covers recent updates in meta-material–based devices for harmless energy harvesting.

Tian *et al.* (2022) [24] frame the utilization of meta-materials for managing the introduction of remote communication systems, including antennas, radomes, and repeat express surfaces. Their survey analyzes the intriguing electromagnetic properties of meta-materials and their impact on communication developments. Chen *et al.* (2021) [7] examine the integration of meta-materials into photonic and optoelectronic devices for applications, for instance, optical modulators, photodetectors, and light-emitting diodes. Their review sets the imperative of meta-materials for achieving reduced areas of strength for circuits.

Tzarouchis *et al.* (2024) [25] focus on the usage of meta-materials for biomedical imaging and therapy, inspecting applications for X-ray, ultrasound imaging, and unsafe improvement treatment. Their outline looks at the surprising properties of meta-materials that contribute to imaging and therapy progress. Ge *et al.* (2024) [11] examine the emerging inspirations driving meta-materials in farming and food production, including sensors for monitoring crop health, packaging for food security, and precision agriculture strategies. Their overview blends the application of meta-materials to address challenges in food security and sustainability.

This writing frame has given an extensive graph of recent advancements in meta-materials research, focusing on their ability to control light and sound waves for high-level prompts. From speculative forward jumps to practical applications across various fields, meta-materials continue to broaden the impediments of what is possible in the space of materials science and engineering. As experts delve further into the properties and potential uses of meta-materials, the future looks promising for the development of innovative technologies that harness the power of these amazing materials.

Research Methodology

The philosophy thoroughly examines existing pieces, articles, web diaries, and discretionary data. This approach means mixing the continuous status of data and seeing key improvements in the field of meta-materials, particularly focusing on their applications in orchestrating light and sound waves for state-of-the-art progress. The most focal stage in the framework is to coordinate a wide-making search using enlightening instructive records, veritable journals, and extraordinary electronic sources. This search is key for gathering immense articles, research papers, and totally capable terminations on meta-materials and their various applications. Moreover, web diaries and discretionary data sources are taught to procure pieces of information from industry-instructed specialists and trained professionals, as well as to see any emerging models or advancements in the field. While the making review is done, the following stage incorporates separating and arranging the amassed information. This affiliation arranges sensing customary subjects, endlessly models across the association, and investigates the sufficiency and authenticity of the sources.

By essentially assessing the ongoing association, the evaluation aims to give a general diagram of the current status of meta-material creative work. Considering the exposures from the making frame, a concluded design is proposed to guide future creative work in the field. This construction is informed by the preferred model, which is evident through the survey cycle, and aims to provide an organized procedure for managing, sorting out, and doing meta-material–based upgrades. The sensible framework outlines key examinations, rules, and strategies for organizing light and sound waves using meta-materials, with a feature that looks for current concerns and researches new pathways in the field. All through the assessment connection, careful frameworks are used to ensure the authenticity and reliability of the disclosures. This includes the joint spread out measures for picking and surveying making, as well as from an overall perspective investigating and unraveling the amassed data. By following a purposeful procedure for managing and keeping an eye on existing association as well as to propose a choosen plan, the assessment wants to add to driving the wisdom and utilization of meta-materials in state-of-the-art upgrades.

Research Findings

Research in meta-materials has made enormous strides lately, offering promising streets for disturbing different mechanical fields. This paper dives into the latest degrees of progress in meta-material arrangement, focusing on controlling light and sound waves to open abilities to outstanding state-of-the-art drives. Meta-materials, portrayed by their astounding development and properties not found in routinely happening materials, obviously stand isolated for their capacity as far as possible constrained by traditional materials. By coordinating designs at the surprisingly unobtrusive and nanoscale, experts have chosen to control the method for managing the acting of electromagnetic and acoustic waves with unmatched exactness Chandrasekhar, A. (2011) [6]. One of the essential revelations in meta-material assessment is the ability to plan materials that show negative refractive records, engaging the bowing of light in capricious ways. This development has been organized to improve super lenses for settling objects past the diffraction farthest reaches of ordinary optics, hence changing imaging and microscopy methodologies.

In this way, meta-materials have been outfitted to make covering devices that can give objects unpretentious to unequivocal frequencies of light, lifting momentous applications in secret progress and optical masks. In acoustics, meta-materials have attracted the control of sound waves in penchants thought unfathomable Li, F. (2020) [17]. Experts have made acoustic meta-materials fit for arranging and focusing sound waves with great limit, impelling developments in acoustic imaging, aggravation control, and cut-down communication. Moreover, meta-material–based acoustic covers have been proposed to free the effects from seismic waves, offering potential applications in quake security and key arrangement. Despite these shocking degrees of progress, a few hardships persist in the field of meta-material planning. One of the major questions is the flexibility of gathering strategies expected to convey meta-materials with dispersed nanoscale structures.

Zadpoor (2016) [33] states that current creation techniques are critical pieces of the time, and they depend on surprising and outrageous cycles, confining the expansive get-together of meta-material developments. Moreover, the introduction of meta-materials is basically sensitive to standard variables like temperature, humidity, and mechanical strain, presenting challenges for dependable applications where power and unwavering quality are indispensable. The general issue declaration driving this assessment is the need to overcome these specific barriers and translate the theoretical limits of meta-materials into practical, flexible solutions for advanced technologies. Addressing these challenges requires an interdisciplinary effort among materials well-informed authorities, physicists, planners, and industry partners to foster innovative methodologies, enhance material execution, and explore new applications across various fields. Yet again, by confronting these challenges directly, meta-material assessment holds the promise of unlocking mechanical development, where controlling light and sound waves opens ways of managing somewhat early astounding possible results.

Recent research has highlighted the use of device learning (ML) algorithms to format meta-materials with tailored properties. ML permits the instant identification of configurations that produce specific desired behaviors in meta-materials. By training models on modern fabric databases, researchers can look for and create new meta-materials structures without resorting to trial-and-error strategies. Studies on nonlinear meta-materials have progressed significantly, particularly in developing materials that interact in a completely unique manner based on the intensity of light or sound waves. These meta-materials offer the potential for tunable devices, permitting applications in dynamic environments where traditional linear materials fail. Furthermore, meta-materials designed using topology optimization techniques have shown excellent improvement in controlling wave propagation. These techniques allow for optimization across multiple scales, resulting in incredible manipulation over light and sound wave interactions at both micro- and macroscopic levels. Hybrid meta-materials, which combine traditional materials with 2D materials like graphene, exhibit superior mechanical, optical, and acoustic properties, paving the way for more robust and flexible applications. Analyzing the results of these hybrid structures on future generations may be explored in extra detail (Vaddadi, 2024) [27].

Meta-materials are being employed to enhance acoustic cloaking devices, which render objects undetectable to sound waves. These cloaks utilize tailored meta-materials to bend sound waves around an object without scattering them, creating an effect of invisibility. The development of sound wave meta-materials in scientific applications, particularly in ultrasound imaging, is gaining significant interest. Research has shown higher imaging talents through the technique of focusing sound waves more effectively through the use of meta-fabric lenses, resulting in higher resolution and better tissue penetration. The function of meta-materials in noise manipulation and vibration suppression is experiencing significant advances. Meta-material–based noise barriers clearly outperform traditional solutions by providing more effective sound dampening with significantly less material. This is due to their ability to control sound waves more efficiently. In architectural acoustics, acoustic meta-surfaces have become valuable for controlling room acoustics by designing surfaces that selectively absorb or diffuse sound (Deng *et al.*, 2024) [10]. Analyzing their potential in smart building generation, where adaptive acoustic environments are required, is a contemporary area of exploration.

Breakthroughs in nanophotonics have leveraged meta-materials to create ultra-thin lenses, known as metalenses, which focus light more precisely than traditional glass lenses. These lenses are set to revolutionize fields such as microscopy, imaging, and customer electronics. Research into chiral meta-materials shows great promise in the area of polarization control, which is essential for communication technology. These materials manipulate the polarization of light in ways that traditional materials cannot, providing new possibilities for optical signal processing. The potential of materials with negative refractive index to create first-class lenses, or "superlenses," marks an outstanding development in the optical domain. These materials enable imaging beyond the diffraction limit, providing new capabilities in areas like bio-imaging, where high-resolution data are essential. Quantum meta-materials, which encompass quantum mechanical principles in their design, represent an emerging frontier. These materials hold the potential for quantum computing and communication technologies, providing new strategies to control quantum states of light for faster and more secure data transmission.

Proposed Conceptual Framework

Table 2.1 Variables of proposed conceptual framework.

Variable	Definition
Material Composition	The relationship of meta-materials, including the course of action and properties of constituent materials, clearly impacts their ability to control light and sound waves. Variations in material synthesis can enable various functionalities and applications in state-of-the-art advances.
Structural Design	The fundamental model of meta-materials, along with their evaluation, periodicity, and perspectives at the nano- and microscale, plays a basic part in picking their electromagnetic and acoustic properties. Developments in central arrangement engage the improvement of meta-materials with exceptionally made wave control limits.
Fabrication Techniques	Improvements in creation systems, including lithography, 3D printing, and self-party techniques, provide precise control over the fabrication of meta-materials. Assortment in systems affects the adaptability, cost-adequacy, and execution of meta-material–based devices and plans.
Wave Interaction Mechanism	Understanding the fundamental science behind wave interaction phenomena, including reverberating, dissipating, and scattering, informs the design and update of meta-material structures. Controlling these interaction phenomena contemplates the improvement of novel functionalities, such as covering, super-resolution imaging, and acoustic focusing.

(*Continued*)

Table 2.1 Variables of proposed conceptual framework. (*Continued*)

Variable	Definition
Environmental Factors	Environmental factors, such as temperature, pressure, and external electromagnetic or acoustic fields, can influence the performance and stability of meta-materials in real-world applications. Taking into account the outcomes of run-of-the-mill conditions is essential for ensuring the reliability and longevity of meta-material–based technologies.
Material Structure Optimization	This variable integrates the optimization strategies used to systematically design meta-materials at the micro- and nanoscales, thereby influencing their electromagnetic and acoustic properties. By mediating the connection between meta-material arrangement and technological advancements, material plan improvement plays a fundamental part in understanding the potential of meta-materials for state-of-the-art levels of progress.
Meta-Material Advancement	It serves as a dependent variable in the research model, assessing the development and progress within the field of meta-materials. It encompasses advancements in material design, manufacturing techniques, and applications, reflecting the impact of various independent factors on expanding the constraints of meta-material science and technology.

Suggestion

Table 2.1 and Figure 2.1 shows the list of variables involved in the proposed Conceptual framework and how each independent variable contributes to the meta-material advancement. To address these challenges, researchers propose creative diagrams throughout the development process of meta-materials. A crucial aspect lies in shifting production techniques to promote versatility and cost-effectiveness. Novel techniques, such as nano-imprint lithography, self-assembly systems, and 3D printing, offer promising avenues for streamlining production cycles and diminishing assembly costs. Also, an interdisciplinary collaboration among material scientists, physicists, originators, and industry stakeholders facilitates the integration of meta-materials into practical devices and applications. By using authority from various fields, experts can oversee complex challenges and accelerate the understanding of meta-material

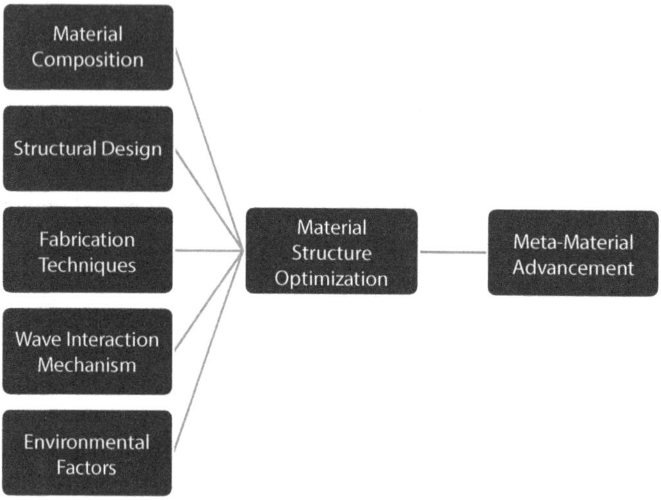

Figure 2.1 Proposed conceptual framework.

evaluation concerning real-world applications. Another significant point is the development of solid arrangement frameworks and simulation tools to improve the performance and functionality of meta-materials (Patel *et al.*, 2022) [20]. Computational appearance methods, such as finite element analysis and fully coupled wave examination, allow experts to predict and manage the behavior of meta-material designs under extra-ideal conditions. These simulation tools support the creation of meta-material–based devices with modified properties, leading to upgraded performance and durability. Likewise, standardization offers a chance that shows meta-materials depiction and testing work with reproducibility and similarity across different assessment survey Krushynska, A. O. (2023) [15]. By setting standards and designs, inspectors can ensure the reliability and validity of experimental results, thus advancing the overall impression of meta-material methods.

The massive use of meta-substances in surveillance and stealth technology is likely to raise moral problems surrounding privacy and security. Governments and organizations need to consider how these substances are utilized in both civilian and military contexts, ensuring that their deployment does not infringe upon individual rights or violate international law. Environmental effect is another vital issue. The manufacturing of advanced meta-substances frequently consists of hazardous chemicals and processes. Therefore, the improvement of environmentally friendly manufacturing strategies should be prioritized to ensure that meta-substances do not contribute to environmental degradation, mainly in large-scale applications (Wang *et al.*, 2024) [30]. Accessibility is also an important area of concern.

As meta-substances revolutionize industries like healthcare, there may be a chance that only wealthy countries and organizations will benefit from these advancements. Discussing suggestions and techniques to ensure equitable access to meta-materials technology should be a significant focus in future research directions.

Additionally, examiners emphasize the significance of exploring new application areas and addressing sensible challenges to advance the commercialization of meta-materials. The undertakings with industry lace attracts experts to see and zero in on application-express necessities, coordinating the approach and improvement of meta-material–based systems. By focusing on high-impact applications such as media communications, sensing, and imaging, examiners can show the benefits of meta-material advances and gain insights from collaborators. Moreover, enlightening initiatives have highlighted the establishment of a remarkable new era for meta-material experts and designers, who play a crucial role in driving progress and advancing the field. By furnishing students with data, the academic world can maintain a burning climate of cutoff and transcendence to drive happened with progress in meta-material imaginative work. Overall, the field of meta-materials offers unimaginable opportunities for organizing light and sound waves, enabling significant advancements (Dalave *et al.*, 2024) [9]. While huge assessment disclosures have shown the significant potential of meta-materials, several challenges persist, from complex assembly arrangement to performance improvement and application integration. Through ongoing efforts and creative plans, researchers can overcome these hardships and unlock the full potential of meta-materials for a broad number of applications, driving technological development and societal impact in the years to come.

Material plan optimization plays a fundamental part in advancing meta-materials, significantly contributing to their extraordinary potential in various fields, such as optics and acoustics, without a doubt. When shaped with evaluations like material creation, hidden away model, produce systems, wave venture instruments, and standard parts, material progression smoothing out changes into an establishment to additionally foster outrageous forefront meta-materials with invigorating functionalities. At its core, material design optimization involves arranging the method and properties of the constituent elements within a meta-material to achieve the required characteristics. This cycle is made by a deep impression of the interaction between material composition and theoretical modeling. By precisely organizing the internal structure, specialists can control how light and sound waves interact with the meta-material, leading to novel theoretical notions like negative refraction, super-resolution imaging, and acoustic covering.

One of the primary challenges in material design optimization is balancing conflicting necessities, such as achieving extraordinary properties (e.g., negative refractive index) while maintaining legitimacy and manufacturability.

Certain computational approaches, including bound part evaluation and assessments, are used to explore huge design spaces and identify ideal plans. This iterative process enables experts to push the restrictions of what is genuinely possible, leading to the advancement of meta-materials. Manufacturing strategies play a focal part in producing refined material designs with both accuracy and flexibility. Whether utilizing nanolithography, 3D printing, self-party, or other procedures, the selected manufacturing approach must be compatible with the highest possible evaluation of key metrics and material properties. Moreover, normal elements like temperature, humidity, and mechanical tension during production should be carefully controlled to ensure reproducibility and quality. The potential for material innovation extends beyond individual meta-material parts to their integration into functional devices and systems. By tailoring the fundamental design of meta-materials to specific application requirements, such as antennas or acoustic meta-materials, experts can unlock their greatest potential in credible circumstances.

This broad perspective surpasses any comparisons between key assessment and practical applications, paving the way for significant advancements in fields ranging from broadcast communications to healthcare. Furthermore, understanding the wave coalition parts within streamlined meta-material designs opens new avenues for controlling light and sound at the nanoscale. By exploiting unconventionalities, such as plasmonics, phononics, and meta-material–induced transparency, analysts can develop innovative wave management techniques that were previously thought impossible. These breakthroughs have broad implications for various interconnected fields, including imaging, sensing, trades, and energy harvesting (Lee *et al.*, 2024) [16]. Overall, material design optimization serves as a fundamental driver for the development of meta-materials, enabling the creation of tailored functionalities that go beyond the restrictions of average materials. By synergistically orchestrating considerations such as material connection, central association, creation strategy, wave cooperation parts, and ordinary components, material advancement speeds up the improvement of state-of-the-art developments that outfit light and sound in astonishing ways.

Conclusion

In the space of an unsurprising turn of events, the evaluation of meta-materials has emerged as a primary field of research, promising to change various endeavors with their extraordinary ability to control light and sound waves. The study, titled "Advances in Meta-Materials: Organizing Light and Sound Waves for Top Tier Levels of Progress," embodies the complexities of these extraordinary materials, uncovering their ability to reshape the area of progress, considering everything. At the point of this assessment is the significant appreciation of meta-materials, which challenge standard material properties by showing organized characteristics not found in naturally occurring substances. Through fussy blueprint and gathering processes, inspectors have gained the capacity to control the behavior of electromagnetic and acoustic waves with unrivaled accuracy. This precise control has opened up numerous applications across various spaces, from media communications and imaging to acoustic covering and energy harvesting. One of the most gigantic pieces of this assessment is its implications for optical and photonic degrees of progress.

By controlling the communication of light using meta-materials, experts have achieved breakthroughs that were once considered vast. From incredibly unstable central habitats prepared for capturing basic standard images to covering devices fit for giving articles invisible to specific frequencies, the logical explanations behind optical meta-materials are extensive. Additionally, the coordination of meta-materials into photonic circuits has opened new avenues in data management and communication, promising faster and more efficient data transfer and evaluation. Similarly, research into meta-materials in the field of acoustics has yielded remarkable outcomes. By orchestrating the provoking of sound waves through incredibly coordinated structures, experts have made novel systems for sound control. Acoustic meta-materials have shown their capacity in applications like noise reduction, sonic covering, and even the redirection of sound waves, paving the way for more specific circumstances and creative sound technologies.

Additionally, the interdisciplinary approach of meta-material assessment has fostered collaboration across various scientific fields, leading to synergistic developments and novel hybrid materials with multifunctional limits. By orchestrating contemplations from physics, materials science, and engineering, presumably, experts are continually pushing the boundaries of what is attainable with meta-materials, expanding the horizons of creative possibility. Overall, the meta-material assessment is poised to significantly affect our understanding of material science and practical applications. By handling the exceptional properties of these organized materials, scientists have unlocked a wealth of opportunities to design and manipulate light and sound waves for state-of-the-art enhancements. As we delve further into the realms of optical and acoustic control, the journey of discovery continues, promising a future where the vast becomes reachable, and the incomprehensible becomes a reality.

References

1. Alù, A., Designer matter: Fascinating interactions of light and sound with metamaterials. *MRS Bull.*, 42, 9, 677–682, 2017.
2. Askari, M., Hutchins, D.A., Thomas, P.J., Astolfi, L., Watson, R.L., Abdi, M., Clare, A.T., Additive manufacturing of metamaterials: A review. *Addit. Manuf.*, 36, 101562, 2020.
3. Barri, K., Jiao, P., Zhang, Q., Chen, J., Wang, Z.L., Alavi, A.H., Multifunctional meta-tribomaterial nanogenerators for energy harvesting and active sensing. *Nano Energy*, 86, 106074, 2021.
4. Boardman, A.D., Grimalsky, V.V., Kivshar, Y.S., Koshevaya, S.V., Lapine, M., Litchinitser, N.M., Shalaev, V.M., Active and tunable metamaterials. *Laser Photonics Rev.*, 5, 2, 287–307, 2011.
5. Cerniauskas, G., Sadia, H., Alam, P., Machine intelligence in metamaterials design: a review. *Oxford Open Mater. Sci.*, 4, 1, itae001, 2024.
6. Chandrasekhar, A., Kaimal, V.P., Bhamare, C., Khosla, S., Ambient intelligence: Next generation technology. *Int. J. Comput. Sci. Eng.*, 3, 6, 2491–2497, 2011.
7. Chen, M., Lu, L., Yu, H., Li, C., Zhao, N., Integration of colloidal quantum dots with photonic structures for optoelectronic and optical devices. *Adv. Sci.*, 8, 18, 2101560, 2021.
8. Dal Poggetto, V.F., Bioinspired acoustic metamaterials: From natural designs to optimized structures. *Front. Mater.*, 10, 1176457, 2023.
9. Dalave, P.A., Joseph, T., Patil, D., *et al.*, Meta-structures for energy absorption: materials, designs, and applications in additive manufacturing and its future scope. *J. Braz. Soc. Mech. Sci. Eng.*, 46, 518, 2024.
10. Deng, Q., Du, T., Gomaa, H., Cheng, Y., An, C., Methods of Manipulation of Acoustic Radiation Using Metamaterials with a Focus on Polymers: Design and Mechanism Insights. *Polymers*, 16, 17, 2405, 2024.
11. Ge, H., Ji, X., Jiang, Y., Wu, X., Li, L., Jia, Z., Zhang, Y., Tri-band and high FOM THz metamaterial absorber for food/agricultural safety sensing applications. *Opt. Commun.*, 554, 130173, 2024.
12. Govindaraman, L.T., Arjunan, A., Baroutaji, A., Robinson, J., Olabi, A.G., Metamaterials for energy harvesting, 2021.
13. Imani, I.M., Kim, H.S., Shin, J., Lee, D.G., Park, J., Vaidya, A., Song, H.C., Advanced Ultrasound Energy Transfer Technologies using Metamaterial Structures. *Adv. Sci.*, 11, 31, 2401494, 2024.

14. Kumar, R. and Mitra, A., Inverse design of dynamically tunable phase-change material based metamaterial absorber induced structural color. *Photonics Nanostruct. Fundam. Appl.*, *54*, 101135, 2023.
15. Krushynska, A.O., Torrent, D., Aragón, A.M., Ardito, R., Bilal, O.R., Bonello, B., Wright, O.B., Emerging topics in nanophononics and elastic, acoustic, and mechanical metamaterials: an overview. *Nanophotonics*, 12, 4, 659–686, 2023.
16. Lee, D., Chen, W., Wang, L., Chan, Y.C., Chen, W., Data-Driven Design for Metamaterials and Multiscale Systems: A Review. *Adv. Mater.*, 36, 8, 2305254, 2024.
17. Li, F. and Hu, R., Metamaterials-enabled sensing for human-machine interfacing. *Sensors*, 21, 1, 161, 2020.
18. Liao, G., Luan, C., Wang, Z., Liu, J., Yao, X., Fu, J., Acoustic metamaterials: A review of theories, structures, fabrication approaches, and applications. *Adv. Mater. Technol.*, 6, 5, 2000787, 2021.
19. Lvov, V.A., Senatov, F.S., Veveris, A.A., Skrybykina, V.A., Díaz Lantada, A., Auxetic metamaterials for biomedical devices: current situation, main challenges, and research trends. *Materials*, 15, 4, 1439, 2022.
20. Patel, V., Chesmore, A., Legner, C.M., Pandey, S., Trends in workplace wearable technologies and connected-worker solutions for next-generation occupational safety, health, and productivity. *Adv. Intell. Syst.*, 4, 1, 2100099, 2022.
21. Santonocito, A., Patrizi, B., Toci, G., Recent advances in tunable metasurfaces and their application in optics. *Nanomaterials*, *13*, 10, 1633, 2023.
22. Simovski, C. and Tretyakov, S., *An introduction to metamaterials and nanophotonics*, Cambridge University Press, University Printing House, Cambridge, United Kingdom, 2020.
23. Suresh Kumar, N., Naidu, K.C.B., Banerjee, P., Anil Babu, T.V.S.R.P., Venkata Shiva Reddy, B., A review on metamaterials for device applications. *Crystals*, 11, 5, 518, 2021.
24. Tian, S., Zhang, X., Wang, X., Han, J., Li, L., Recent advances in metamaterials for simultaneous wireless information and power transmission. *Nanophotonics*, 11, 9, 1697–1723, 2022.
25. Tzarouchis, D.C., Koutsoupidou, M., Sotiriou, I., Dovelos, K., Rompolas, D., Kosmas, P., Electromagnetic metamaterials for biomedical applications: short review and trends. *EPJ Appl. Metamater.*, 11, 7, 2024.
26. Uriri, S., Ismail, Y., Petruccione, F., Quantum metamaterials: applications in quantum information science, 2020, *arXiv preprint arXiv:2006.03757.*
27. Vaddadi, K., Beyond AI in Pushpaka Vimana: Ancient Indian Aerospace Technology and Modern Space Science. *Acceleron. Aerosp. J.*, 3, 2, 474–478, 2024.
28. Valipour, A., Kargozarfard, M.H., Rakhshi, M., Yaghootian, A., Sedighi, H.M., Metamaterials and their applications: an overview. *Proceedings of the Institution of Mechanical Engineers, Part L: Journal of Materials: Design and Applications*, vol. 236, pp. 2171–2210, 2022.

29. Wang, J., Dai, G., Huang, J., Thermal metamaterial: Fundamental, application, and outlook. *Iscience*, 23, 10, 1, 2020.
30. Wang, Y., Sha, W., Xiao, M., Gao, L., Thermal Metamaterials with Configurable Mechanical Properties. *Adv. Sci.*, 11, 40, 2406116, 2024.
31. Wu, B., Jiang, W., Jiang, J., Zhao, Z., Tang, Y., Zhou, W., Chen, W., Wave Manipulation in Intelligent Metamaterials: Recent Progress and Prospects. *Adv. Funct. Mater.*, 2316745, 1, 2024.
32. Wu, K., Duan, G., Zhao, X., Chen, C., Anderson, S.W., Zhang, X., Metamaterial-enhanced near-field readout platform for passive microsensor tags. *Microsyst. Nanoeng.*, *8*, 1, 28, 2022.
33. Zadpoor, A.A., Mechanical meta-materials. *Mater. Horiz.*, 3, 5, 371–381, 2016.
34. Zhang, J., Hu, B., Wang, S., Review and perspective on acoustic metamaterials: From fundamentals to applications. *Appl. Phys. Lett.*, 123, 1, 1, 2023.
35. Zhong, J., Zhao, C., Liu, Y., Ren, J., Yang, C., Zhang, Z., Meta-materials of Re-entrant Negative Poisson's Ratio Structures Made from Fiber-Reinforced Plastics: A Short Review. *Fibers Polym.*, 25, 2, 395–406, 2024.

3

Advanced Intersection of Material and Medicine Revolutionizing Healthcare Outcomes

Ashwini A.* **and Ferlin Deva Shahila**

Department of Electronics and Communication Engineering, Vel Tech Rangarajan Dr. Sagunthala R&D Institute of Science and Technology, Chennai, Tamil Nadu, India

Abstract

The convergence of material science and medicine is an important turning point in healthcare, ushered in an era of invention with transformative consequences. This trans-disciplinary field, known as the "Advanced Intersection of Material and Medicine," encompasses an extensive range of novel findings as well as applications that are changing the face of healthcare delivery. The fields of nanotechnology and material studies are inextricably linked, with nanotechnology providing gadgets, approaches, and methods for influencing content at the nanoscale, while material science focuses on comprehending and utilizing the unique properties of these constructed substances for use in beneficial ways. From nanotechnology-enabled medicine delivery methods to 3D-printed organs, the joint integration of cutting-edge substances and healthcare techniques is revolutionizing treatment for patients in a wide range of areas. One of the major difficulties in the area is the development of nanoparticles for customized delivery of drugs, allowing for accurate medicinal doses while reducing side effects. Additionally, survival nanoparticles facilitate the development of medical devices and technologies that communicate easily with biological systems, thereby decreasing rejection rates and enhancing patient outcomes. Furthermore, improvements in additive manufacturing and bio-printing technology enable the creation of tailored devices and cells, resulting in unprecedented levels of customization and potency in treatment alternatives. Additionally, intelligent materials and cell phones with sensors

**Corresponding author*: a.aswiniur@gmail.com

Kolla Bhanu Prakash, S. V. Ranganayakulu and K. S. Jagannatha Rao (eds.) New Frontiers in Materials Science: Interdisciplinary Approaches to Innovation and Technologies, (37–60) © 2025 Scrivener Publishing LLC

provide patients and medical staff with continuous evaluation capacity, enabling earlier disease detection and personalized approaches to therapy. Regenerative surgery and tissue science, powered by innovative framework components and treatments using stem cells, offer the ability to restore destroyed organs and tissues, revolutionizing the medical management of severe illnesses. The targeted drug delivery gives an efficiency of 90%, which shows the improvement in integrating the material in the medical field. It is clear that the complex intersection of materials and medical sciences will soon transform the limits of medical care, driving toward an age in which enhanced patient outcomes and a higher standard of existence are no longer desires but achieved truths.

Keywords: Artificial intelligence, augmented reality, block chain, digital healthcare, deep learning, machine learning, metahealth, metaverse

3.1 Overview of Nanotechnology

Nanotechnology has radically revolutionized the field of material science by giving techniques and tools for creating materials at a nanoscale, resulting in a wide range of novel uses. At the heart of this shift is the capacity to precisely alter matter at sizes ranging from 1 to 100 nanometers. Such extraordinary ability to regulate material properties permits the creation of nanostructures with superior physical, electrical, thermal energy, and visual qualities. Carbon nanotube- and graphene-based substances, for example, are very strong and conductive, thus being interesting possibilities for contemporary composite and electronic devices. Furthermore, the nanotechnology process is a distinct quantum that helps in creating materials with new functions, like quantum dots, that are miniature. These act as very effective diodes that help in producing light for sensitive-based identification [1].

Furthermore, the nanoscale helps in the modification of surfaces that are made to vanish for specific purposes. It helps in group attachment, which functions for the biological molecules or size of the particles. This results in qualities related to catalysis and biological functions. They possess different uses such as absorption of water and its filtration, which helps prevent rust. The nanoscale helps in the creation of nanocatalysts, which have large surface areas with chemical characteristics. These are used in various industries including chemical production, oil refining concepts, and environmental clean-up, enhancing the efficiency with the economic factors. These nanostructures help drive the processes more effectively by producing substances with peculiar characteristics for a wide range of purposes.

3.1.1 Nanotechnology's Impact on Material Sciences

Nanotechnology has created a transformation in material sciences which helps in controlling and designing nanoscale substances within the range of 1 to 100 nanometers [2]. This discipline provides unparalleled oversight of substances' order, characteristics, and efficiency, resulting in a wide range of uses across sectors. Here's a summary of nanotechnology's influence on physical research:

3.1.1.1 Nanomaterials Synthesis

Nanoparticles enable the production of nanostructures with specific characteristics using a variety of processes, including bottom-up (e.g., chemically vaporized evaporation, sol–gel production) and top-down (e.g., printing, mechanical grinding) techniques. These techniques permit the creation of particles, tiny wires, and nanotubes, which are tiny films with exact control over their shape, size, and content [3].

3.1.1.2 Enhanced Mechanical Properties

Nanostructures have resulted in the production of substances with superior mechanical properties, such as increased strength, resilience, and elasticity. Nanotube- and graphene-reinforced composites, for example, have high toughness, which makes them perfect for use in aircraft, automobiles, and building materials.

3.1.1.3 Improved Electrical and Thermal Conductivity

Nanomaterials, including nanoparticles of metal, quantum dot particles, and materials made from carbon, have distinct electric and thermodynamic characteristics [4]. They are capable of great conductivity, making them useful in electronics, detectors, and battery systems. Furthermore, materials such as nanowires and nanocomposites have high thermal conductivity, which makes them ideal for regulating temperature applications.

3.1.1.4 Surface Modification and Functionalization

The use of nanotechnology enables precise control over the surface characteristics of substances, resulting in increased functionality. Surface modification methods such as chemical modification, treatment with plasma, and self-assembly allow for the attachment of biomolecules, catalysts, as well as other functional groups, conferring specialized qualities such as biocompatibility, catalytic capacity, and self-cleaning interfaces.

3.1.1.5 *Advanced Coatings and Films*

Nanoparticles enable the creation of thin coatings and films with specific qualities, such as visual transparency, resistance to scratches, and barrier properties [5]. Nanostructure material coatings helps in preventing rust and degradation from the environment, boosting the lifespan of objects under extreme circumstances.

3.1.1.6 *Nanostructured Catalysts*

Nanomaterials are effective catalysts due to their large surface area and unusual chemical properties. Nanocatalysts have increased their catalytic activity and specificity, which leads to higher effectiveness in chemical reactions including hydrogenation, burning, and fuel cell operation. They have uses in cleaning up the environment, manufacturing drugs, and conversion of energy activities.

3.1.1.7 *Smart and Responsive Materials*

Nanoparticles permit the creation of smart substances that react to environmental inputs like heat, light, and acidity. Nanomaterial-based actuators and sensors respond quickly and sensitively, making them suitable for uses such as medical care, surveillance of the environment, and automation.

3.2 Nanomaterials for Imaging and Diagnosis

In addition to imaging, nanoparticles can help improve diagnosis precision by developing diagnostics at the point of care and biosensors. Nanoparticle-based biological sensors identify markers linked to a variety of illnesses, including infections, heart disease, and degenerative ailments, in a timely and highly sensitive manner [6]. These compact and inexpensive diagnostic instruments allow early illness identification, therapy response tracking, and screening of populations at risk in resource-constrained situations, therefore boosting accessibility to prompt medical care. Figure 3.1 shows the impact of nanomaterials in imaging and diagnosis. This primarily depends on the formulation of nanoparticles.

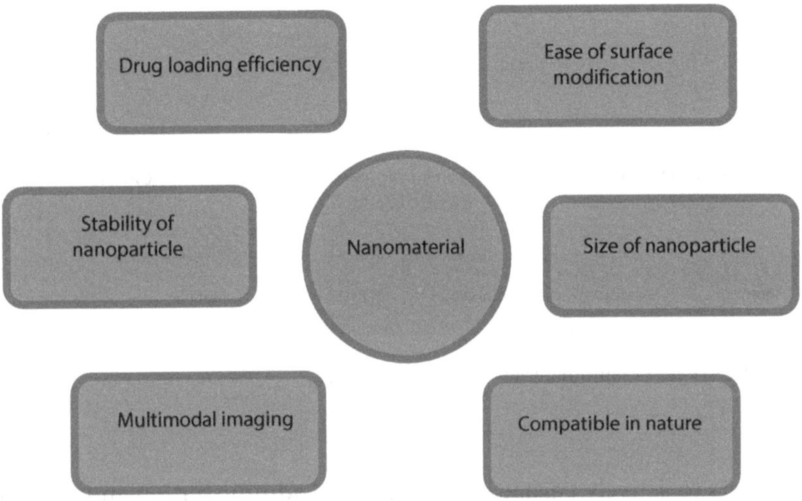

Figure 3.1 Impact of nanomaterials for diagnosis and imaging.

Overall, the combination of nanoscale and healthcare has resulted in major advances in scanning and diagnostics, allowing for quick identification, comprehensive characterization, and individualized management of diseases. Physicians can achieve superior outcomes for patients, less expensive medical care, and a higher quality lifespan for people all over the globe by using nanotechnology's special properties [7].

3.3 Biomaterials and Bioactive Devices for Medical Devices

The confluence of materials and interactive technologies with healthcare has resulted in a shift in healthcare results, notably via the creation of sophisticated medical equipment. Composites are synthetic substances that communicate with living things, whereas active devices are implanted or worn devices that cause specific biological reactions to assist healing of tissues, medication administration, or physiologic monitoring [8]. These novel techniques have greatly enhanced the treatment of patients by providing alternatives that can be more biocompatible, long-lasting, as well as successful than standard medical equipment.

Wearable biosensors, such as smart patches and wearing gadgets, monitor health data, blood sugar levels, and biochemical data while giving ongoing feedback to both patients and medical professionals. These gadgets

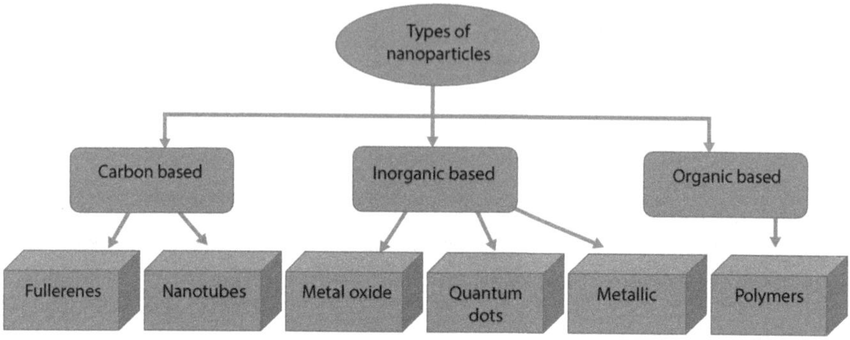

Figure 3.2 Types of nanomaterials in medical devices.

help users to actively engage in their medical administration and allow for online tracking of chronic diseases and telehealth care. Overall, the incorporation of nanomaterials and active technologies into medical technology has transformed health results by providing customized options that match individuals' specific demands [9]. These improvements not only increase the effectiveness of treatment and ease for patients but also help to lower healthcare costs and improve overall quality of life. As the study and creation in this subject progresses, the prospect seems promising for additional inventive ways that will alter the way healthcare is provided overall handling of patients. Figure 3.2 shows the types of bio-nanomaterials and their classification.

3.3.1 Types of Biomaterials and Bioactive Devices

Biomaterials and bioactive devices are utilized in contemporary medical equipment in a variety of substances and capabilities, each one customized to a particular application in health.

3.3.1.1 Biodegradable Polymers

Biodegradable substances like polylactic acid (PLA), polyglycolic acid (PGA), and their mixtures (for example, PLGA) are widely employed in medical equipment [10]. They disintegrate naturally in the human body, rendering them ideal for uses such as stitches, drug routes of administration, and neural templates.

3.3.1.2 *Beneficial Windows and Tiles*

Glasses and ceramics can bind with cells, facilitating the creation of bones and restoration. They are used in dental bridges, bone transplants, and implant surgery to improve bone integration and regenerate tissue [11]. Metal compounds, including titanium, stainless steel, and cobalt–chromium alloys, are frequently used in medical devices due to their strength and biological compatibility. The aforementioned substances are used in orthopedic implants, cardiac stents, and orthodontic appliances.

Hydrogels are water-swollen systems of polymers that resemble the outer layer of cells. Their substantial water content as well as biocompatibility make them suitable for application as surgical dressings, lens coatings, and drug packaging.

3.3.1.3 *Intelligent Polymers*

Intelligent polymers, also referred to as stimulus-responsive polymers, exhibit reversible alterations in response to external stimuli such as temperatures, pH levels, or light. They are used in systems for drug delivery, biosensors, and transplantation to produce regulated pharmacological

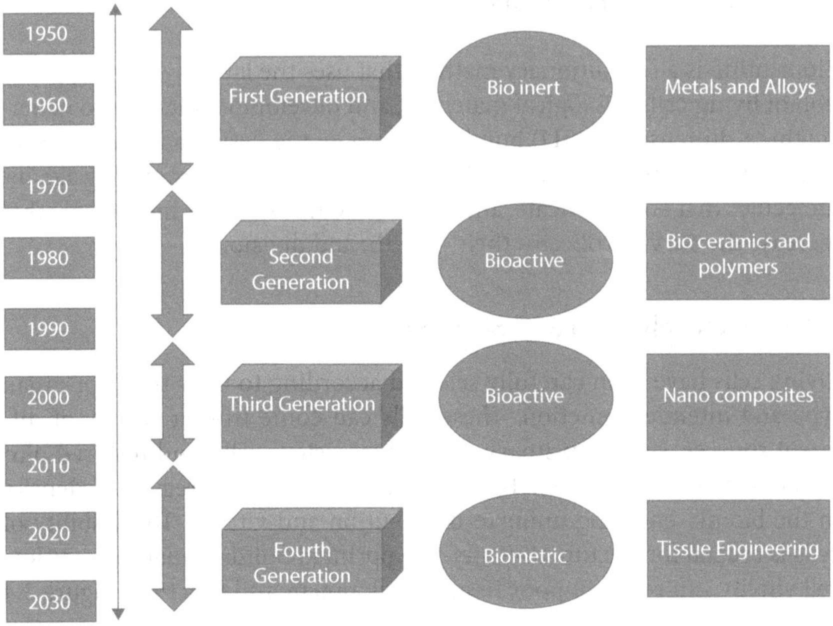

Figure 3.3 Development of material nanosciences.

discharge or to elicit particular reactions in the body. Nanomaterials, including tiny particles, nanofibers, and quantum dots, are used in a variety of medical equipment for images, medication administration, and the creation of tissues [12]. They have unique qualities such as large surface area, adjustable optical characteristics, and individualized administration, making them particularly attractive for nanomedicine. Figure 3.3 shows the development of biomaterials across different generations.

3.3.1.4 Synthetic Polymers

These substances are intended to replicate the framework and functioning of living tissue or organs. For instance, artificial scaffolding for tissue creation, artificial tissues, and biohybrid technologies that integrate living cells. Removable sensors and circuits use biocompatible materials to keep track of vital signs or provide therapeutic stimulation. Heart rate monitors, neuro-stimulators, and glycemic trackers are among the gadgets that help patients and doctors manage their diseases better. These biological materials and active devices are just some examples of the wide variety of substances and capabilities found in sophisticated medical equipment.

3.4 Bioprinting of Living Tissues and Organs

Bioprinting is a revolutionary method that uses the layer-by-layer application of living cells, biological materials, and development factors to generate three-dimensional (3D) builds that imitate the design and functions of real tissue and organs [13]. Bioinks may contain growth-promoting agents, molecules that communicate, and other biologically active ingredients that increase the survival of cells, their growth, and division.

3.4.1 Cell Selection and Seeding

Living cells have been carefully selected according to their tissue or organ type and intended function. These cells can come from a variety of different sources, such as primary cell cultures, stem cells, and induced pluripotent stem cells (iPSCs). The cells are suspended at a regulated density in the bioink, ensuring uniform distribution and vitality. The viability of cells is crucial for the functionality of bioprinted building materials, as low cell vitality can result in poor tissue development and function. Figure 3.4 shows the bioprinting stages of various tissues and organs.

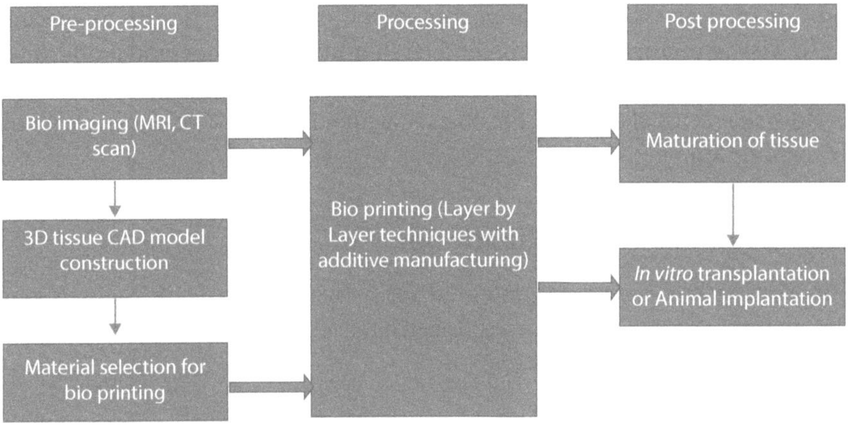

Figure 3.4 Bioprinting process of living tissues and organs.

3.4.2 Bioprinting Process

Bioprinting involves the exact deposit of bioink layers in a geographically regulated method to create sophisticated 3D structures. Various bioprinting processes are used, including inkjet printing, extrusion, and light bioprinting technology. Extrusion-based bioprinting, which uses hydraulic or mechanical pulls to push out bioink via a sprayer, is one of the oldest and most common methods due to its adaptability and capacity for handling enormous concentrations of cells [14].

3.4.3 Post-Printing Processing

After bioprinting, the structure undergoes post-printing treatment to improve cell viability, maturation, and tissue growth. This may include linking the bioink to improve its strength, as well as growing in a bioreactor to provide biological signals like oxygen, nutrition, and stimulation from movement. The bioreactor conditions for cell growth may be tuned to replicate the natural organ milieu, promoting muscle growth and functioning throughout the years.

3.4.4 Tissue Maturation and Integration

Following post-printing the process, the bioprinted structure undergoes tissue maturation and integration, in which cells multiply, differentiate, and organize themselves to mimic the shape or functioning of natural tissues and organs. This procedure might take a few weeks or months, based on the intricacy and measurements of the bioprinted tissues or organs [15]. Tissue maturation can be accelerated by adding hormonal and biomechanical stimuli. It can also be combined with supportive cells like endothelial cells, also called fibroblasts, which increase circulation and cell incorporation.

3.4.5 Applications of Bioprinting

Bioprinting has enormous potential for a variety of uses in the fields of regenerative medicine, tissue engineering, and the development of drugs. Some of the significant uses are:

3.4.5.1 Bioprinted Organs

Bioprinted organs and tissues can be transplanted to substitute destroyed or ill tissues, providing personalized and regenerative therapy for patients suffering from organ disease or loss. In bioprinting, the printer produces several layers of bioink, including cells and substances, by a specified digital design that comes from medical scan images or computer-aided design (CAD) equipment [16]. Bioprinted tissue and organs provide physiologically accurate models for researching disease processes, therapeutic effectiveness, and safety in a regulated laboratory context, eliminating the requirement for laboratory animals and optimizing medicinal product pathways.

3.4.5.2 Individualized Medicine

Bioprinting allows for the creation of patient-specific organs and tissues based on anatomical and physiologic features, providing individualized therapy choices for patients with congenital defects, serious wounds, or degenerative diseases.

3.4.5.3 Biological Studies and Development

Bioprinted tissue and organs are useful instruments for researching basic biological processes, tissue growth, and organogenesis, thereby improving the understanding of intricate ecosystems and cancer physiology.

In conclusion, bioprinting has a chance to transform healthcare by allowing the creation of working organs and tissues for transplantation, disease modeling, customized therapy, and biological research [17]. Ongoing advancements in biological printing is likely to promote more creative thinking and conversion of printed constructions into usage in medicine.

3.5 Smart Fabrics for Health Monitoring

Smart textiles, commonly referred to as e-textiles or electronic clothing, are fabrics that are embedded with electrical parts and sensors to perform a variety of functions, like tracking one's health [18]. These materials blend the softness and adaptability found in conventional textiles with the abilities of contemporary technology to provide a constant and unnoticeable assessment of metabolic and physiological variables.

3.5.1 Sensor Integration

Smart fabrics include tiny sensors that can identify and track a variety of health indicators such as respiration rate, heart rate, body temperature, arterial pressure, and mobility [19]. These sensors can be knitted, manufactured, or stitched right into the material, thus providing continuous and

Table 3.1 Broad classification of smart textile-based sensors.

Smart textile- sensors		
Environmental monitoring sensors	**Health monitoring sensors**	
Sensor based on stimuli exposure	*In vitro* biomonitoring	Physiological sensors
Global positioning systems	Diabetes monitoring	Electrocardiography
Temperature	Uric acid (wound level)	Body temperature
Exposure to UV radiation	Lactate (Wound healing)	Pulse Oximeter
Gas	Bacteria (Wound infection)	Motion
Pressure	Antimicrobial Peptides	Respiration
Humidity	Protease (Wound healing)	Electroencephalography

simple tracking of the wearer's medical state. Table 3.1 shows the smart textile-based sensor monitoring.

3.5.2 Data Collection and Transmission

Sensors that are incorporated in smart textiles capture current data regarding the wearer's health and physiological variables. That data is then analyzed and remotely sent to a cell phone, wristwatch, or other kind of gadget for evaluation and interpretation. Bluetooth and cell phones provide good information transfer, distance surveillance, and immediate valued response [20].

3.5.3 Comfort and Wearability

The interesting and innovative fabrics are breathable, lightweight, and easy to wear, which provides health surveillance even at regular times. They are adaptable to all elastic electronic components, ensuring they accommodate the features of the body, including limited mobility and pain, thus encouraging user adoption.

3.5.4 Biometric Authentication

Biometric identity with authentication helps in tracking health. This is made possible by intelligent textile systems. These devices are integrated into the textiles and can identify distinct patterns with physiological characteristics like handprints, venous patterns, and fingerprints which provide safer management with verification.

3.5.5 Applications

Smart textiles help in health monitoring for walking, exercise, and overall well-being. They are also used in various applications, including sports, medical items, and portable items such as footwear, tops, gloves, and headgear. These smart textiles provide notifications and set goals for various physiological situations or illnesses in natural aspects.

3.5.6 Remote Patient Monitoring

Innovative fabrics let healthcare workers watch patients remotely, enabling them to observe vital signs and metabolism. The use of this technology allows for rapid detection of diseases, timely treatment, and tailored therapy, particularly for persons with chronic ailments, older people, or individuals recovering from surgery or disease [21].

Overall, clever fabrics have a chance to serve as a basis for ongoing and invisible health monitoring, enabling individuals to be in control of whatever is most relevant. As advances in technology occur, smart garments are anticipated to become more complex, with greater sensors and functions to provide detailed and personalized wellness tracking alternatives.

3.6 Brain-Computer Interfaces for Communication and Control

Interfaces for brains constitute an innovative confluence of medical equipment that has altered the way medicine works by providing novel ways of involvement and administration, particularly for persons without serious impairments or neurological illnesses. Brain-computer interfaces (BCIs) provide a direct connection with the brain system and external items, letting individuals interact with their environment, control physical nature, or recover lost vision or motor skills. Here are what medical instruments are transforming the health section:

3.6.1 Neural Recording and Decoding

BCIs record impulses of electricity from the cortex of the brain using contemporary technology and substances, often through strategically placed electrodes or nonsurgical methods such as electroencephalogram (EEG). These neural signals are analyzed and turned into directions that may be used to control outside objects such as laptops, robotic legs, and assistive equipment. The biological compatibility implantation and signal interpretation methods are critical for successfully and consistently obtaining and analyzing neurological signals.

3.6.2 Assistive Communication

BCIs serve as a lifeline to individuals with severe communication difficulties such as locked-in syndrome or late-stage amyotrophic lateral sclerosis (ALS). BCIs enable such individuals to engage with other people through text, artificial voice, or manipulation of assistive technology by reading brain signals connected with speech or movements. Those who might ordinarily be hesitant to express their opinions and wishes can profit from this cutting-edge innovation, which restores their autonomy and sovereignty [22].

3.6.3 Motor Restoration

BCIs offer an opportunity to help those who have been paralyzed or lost a leg restore their ability to move around. BCIs can control limb replacements, exoskeletons, and robotics with incredible precision and precision by interpreting electrical signals linked to movement goals. These neural components help in forming innovative connections that enable the smooth coupling of the user's goals with external machines, which helps individuals do tedious tasks independently.

3.6.4 Neurorehabilitation

BCIs play a key role in the rehabilitation process by providing spontaneous feedback and various modes of exercise, which provide growth after neural damage. They monitor activities during entire therapy sessions and help in adjusting the patient's physical capabilities. This restores the recovery outcomes and helps in promoting neural reorganization.

3.6.5 Cognitive Enhancement

The investigation is currently being conducted on BCIs for brain enhancement exercises such as recall supplementation and attention management. Neural activity associated with particular cognitive processes is monitored and modified by BCIs, which can improve the recall of information, focus, and academic capacity. These programs may aid individuals who suffer from Alzheimer's disease, Parkinson's disease, severe injury to the brain, or cognitive impairment caused by aging.

A remarkable success at the intersection of material and clinical studies is represented by BCIs, offering new paths for contact monitoring and recovery in medicine. As technologies and materials develop, BCIs have the opportunity to enhance the existence of persons with brain diseases and limitations by helping them to move around more fully and actively participate in society. Electronically energized compounds, which respond to electricity by distorting or changing their mechanical characteristics, are considered to have a high chance of producing cells with enhanced functionality and responsiveness. These substances form a growing structure that controls cell turnover, encourages tissue healing, and affects biological events.

3.6.6 Cellular Alignment and Guidance

Encouragement of cell symmetry and guidance may be provided by electrically charged chemicals through the release of electrical signals. The enhancement of cell migration, initiation, and expansion along the line of the electromagnetic field, resulting in matching tissue topologies, is caused by stimulation with electricity. This feature is extremely useful for the construction of anisotropic tissue such as muscles of the heart, skeletal muscle, and nerve cells [23].

3.6.7 Enhanced Cell Proliferation and Differentiation

Cell development and differentiation can be affected by stimulation with voltage from electrically charged chemicals, in addition to the development of tissues and growth. Electrical stimulation can cause embryonic cells to divide into certain generations, like bone marrow stem cells or other cells, resulting in the formation of functional tissues with specific characteristics.

3.6.8 Sensing and Monitoring

Physically charged compounds can operate as biosensors, measuring modifications in electric conductivity, impedance, or capacitance in responses to chemical reactions throughout cells, the functioning of tissues, or environmental stimuli. This capability enables non-invasive study of cell welfare, activity, and response to medication.

Their capacity to influence cellular activity, stimulate regeneration of tissue, and interact with biological systems offers promise for the further development of tissue science and regenerative therapies. As research advances, electrically conductive substances are likely to play an increasingly essential role in the creation of the future of tissue constructions for applications.

3.7 Micro and Nanoscale Carriers for Drug Transport

Micro and nanoscale carriers are emerging as potent instruments at the crossroads between substance medicine and science, transforming medical results by allowing more tailored medication delivery, increased effectiveness in therapy, and fewer side effects. The aforementioned carriers are intended for the packaging and delivery of medicinal molecules, such as medicines, genes, or proteins, to specific locations within the body's tissues [24]. Here's a list of how microscopic and nanoscale carriers are changing medicine delivery and the field of medicine:

3.7.1 Improved Drug Solubility and Stability

Micro and nanoscale carriers can store hydrophobic or fragile pharmaceuticals within their cores, preventing degradation and increasing absorption in physiological fluids. This characteristic is especially useful for medications that are weakly soluble in water since it allows for their distribution in an ongoing and accessible state.

3.7.2 Targeted Drug Delivery

Micro and nanoscale carriers can be designed to target particular tissues, cells, or organs in the body, minimizing adverse reactions while increasing therapeutic efficacy. The coating of transporters with target ligands such as antigens and peptides, permits preferential attachment to receptors or markers overproduced on sick cells, allowing for precise delivery of medications to sick areas. Figure 3.5 shows various nanomaterial-based biomedical applications [25].

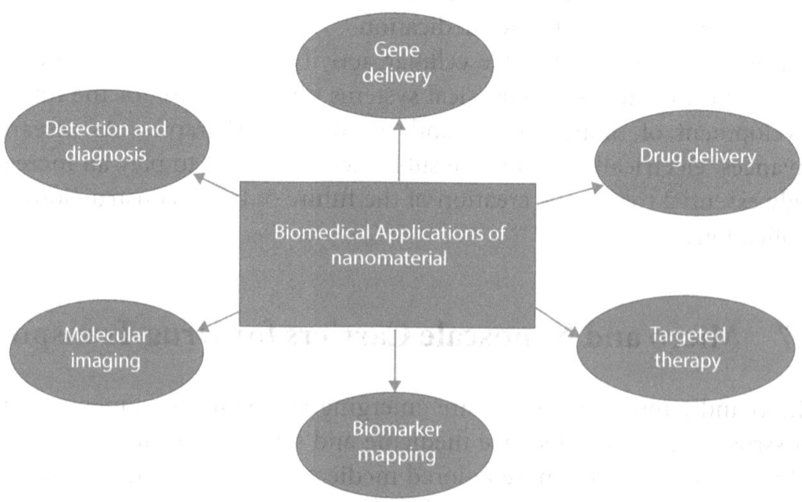

Figure 3.5 Various biomedical applications of nanomaterials.

3.7.3 Controlled Drug Release

Micro and nanoscale carriers can be constructed to disseminate drugs for therapy in a regulated and maintained way, carriers medication access at the intended site and lowering dose frequently. Carrier features such as shape, form, structure, and charge on the surface can be adjusted to control the process of drug release and improve the therapeutic benefit.

3.7.4 Enhanced Cellular Uptake and Intracellular Delivery

Micro and nanoscale carriers can improve the absorption and intracellular delivery of beneficial medicines by bypassing biological obstacles such as cell membranes and lysosome breakdown. Transport features like dimension, the charge on the surface, and functionalization of the surface impact relationships with cells, allowing for effective payload uptake and releases among cells of the target.

3.7.5 Multifunctional Carriers

Micro and nanoscale materials are designed to meet the properties that allow for administrations such as agents and probe diagnostics. This novelty integrated the tools in one platform which can be effectively tailored and assumed for medical treatments.

3.7.6 Minimized Side Effects and Toxicity

Micro and nanocarriers provide a multifunctional range of properties that enable continuous observations of the medical imaging representatives. They combine into one system allowing for illness treatment [26].

3.8 Biocompatible Surgical Instruments and Implants

Implantation with environment-friendly equipment plays an important node between nanomaterials and material sciences. This provides advancement in healthcare which helps in patient safety, reduction in problems, and effective treatment. They come in contact with the biological tissues which induces unpleasant responses. This allows effective operation and long-term improvement in patient care. Here's how biodegradable surgical equipment and implants are transforming medical results.

3.8.1 Enhanced Biocompatibility

Surgical equipment and implants are made from biocompatible materials including titanium, stainless steel, cobalt-chromium metals, and medical-grade plastics. The aforementioned substances are highly biocompatible, reducing the possibility of negative reactions, inflammatory processes, or rejections when in contact with living organisms or liquids. This promotes the safety of patients and lowers the risk of issues after surgery.

3.8.2 Reduced Risk of Infection

Biocompatible substances are frequently developed to have antibacterial qualities, lowering the likelihood of surgery site diseases and implant-related illnesses. Surface changes, like antimicrobial paints or therapies, reduce bacteria adhesion and the creation of biofilm on surgical equipment and implants, resulting in quicker healing and less antibiotic medication.

3.8.3 Customization and Personalization

Developments in materials science and manufacturing technology allow for the personalization and customization of surgical equipment and devices to match each individual's specific morphological and physiologic needs. The use of additive manufacturing (3D printing) technologies enables the creation of implants tailored for patients with intricate geometries and customized mechanical characteristics, thereby enhancing prosthetic fit, order, and usefulness.

3.8.4 Biodegradability and Restorability

Biologically compatible substances can be made reversible or restorable, which means they will progressively degrade and integrate with the tissue around them over time. Biodegradable implants remove the need for removal procedures, which reduces patient suffering and healthcare expenses. They are especially useful for transitory implants like fracture repair devices and drug-eluting matrices that regenerate tissue [27].

3.8.5 Promotion of Tissue Integration and Regeneration

Biocompatible materials aid in tissue incorporation and regrowth, hastening the healing phase and increasing functional recovery. Surface alterations and coating can encourage cell adherence, proliferation, and division on implants, thereby improving tissue growth and incorporation. This feature is essential to the effective functioning of implants in orthopedics, implant dentistry, and the healing of soft tissue devices.

3.8.6 Compatibility with Imaging and Diagnostic Technologies

Biocompatible polymers are suitable with a variety of diagnostic and testing modalities, allowing for precise viewing alongside evaluation of surgical results. Interventions composed of platinum or cobalt-chromium alloys are transparent, enabling easy viewing of X-rays, computerized tomography (CT) scans, and magnetic resonance imaging (MRI), which helps in following surgery evaluation and follow-up visits [28].

3.8.7 Innovations in Minimally Invasive Surgery

Biocompatible polymers have allowed for breakthroughs in less invasive surgical techniques like endoscopy and robotic-assisted operation. Compressed equipment and implants made from biocompatible materials enable doctors to execute challenging operations with more accuracy, less tissue stress, and shorter recovery times, resulting in better results and increased comfort for patients. Table 3.2 shows the comparative results of various fields with the advances and their impact.

Table 3.2 The comparative results of various fields with the advances and their impact.

Technique	Impact	Advances
Nanotechnology	Earlier disease diagnosis	Targeted delivery of drugs
Smart Drug Delivery	Side effect reduction with controlled dosage	Controlled delivery of drugs
Biomaterials	Improved patient recovery rates with a decreased rate of rejection	Implants with biocompatible materials
3D Bio-printing	Reduced time with control of overdosage	Customized organs and tissue printing
Wearable Devices	Real-time disease management	Sensor development

Looking ahead, the prospects at the junction of materials science and medicine are quite promising. Machine learning, AI, and regenerative medicine are among the latest technologies that have the potential to drive advances in healthcare. Advanced materials, such as stimulus adaptability, self-healing abilities, and programmable features, will allow for the creation of intelligent therapies, wearable gadgets, and tissue-engineered constructs with so far unheard-of possibilities. Furthermore, the incorporation of health information technology, telemedicine, and remote patient tracking will allow for more affordable and patient-centered healthcare

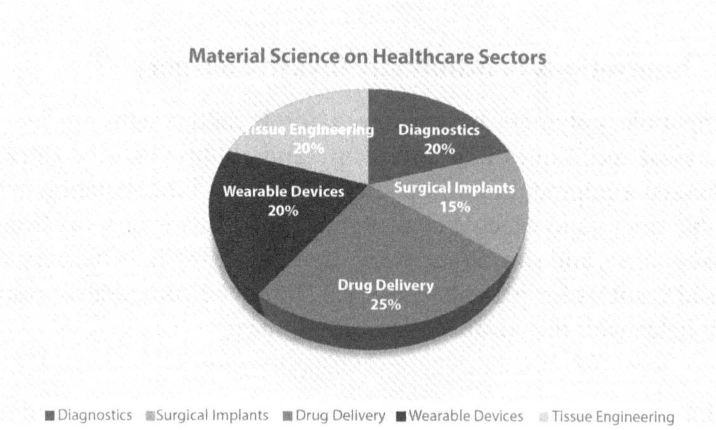

Figure 3.6 The percentage of each healthcare sector and its impact on material science.

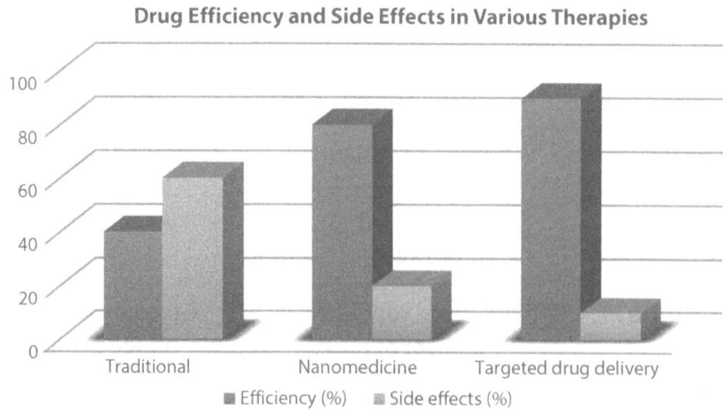

Figure 3.7 Comparative results of various therapies with drug efficiency and side effects.

models. Figure 3.6 shows the percentage of each healthcare sector and its impact on material science.

The intersection between material science, biological processes, and technology continues to encourage multidisciplinary cooperation and promote synergies between learning, commercialization, and clinical research. Figure 3.7 shows the comparative results of various therapies based on efficiency and side effects.

In the future years, overcoming obstacles such as approval from regulators, scalability, and cost will be critical for the broad adoption and transfer of innovative materials and advances into clinical practice.

3.9 Conclusion and Future Directions

To summarize, the combination of the medical and scientific modes has resulted in an interesting discovery that helps in the transformation of health outcomes in various areas. This form of new biomaterial creation incorporates active targets which have led to innovations in the field of bio-printing interfaces, leading to collaboration with healthcare and content-based treatment where the patients are treated effectively with interventions. These provide prominent therapies focused on individual patient needs, leading to increased efficiency and better patient outcomes. The most prominent role is played by biocompatible nanomaterials, which are concentrated on tools, medical devices, remote patient care, problem reduction, and tissue growth. The relationship is vibrant and disciplined, transforming healthcare, patient well-being, and tackling health issues. Thus, proper adoption, teamwork, and medical treatment contribute to the health background and bring a revolution in the medical field.

References

1. Kasula, B.Y., Revolutionizing Healthcare Delivery: Innovations and Challenges in Supply Chain Management for Improved Patient Care. *Trans. Latest Trends Health Sect.*, *15*, 15, 2023.
2. Gidiagba, J.O., Daraojimba, C., Ofonagoro, K.A., Eyo-Udo, N.L., Egbokhaebho, B.A., Ogunjobi, O.A., Banso, A.A., Economic impacts and innovations in materials science: a holistic exploration of nanotechnology and advanced materials. *Eng. Sci. Technol. J.*, *4*, 3, 84–100, 2023.
3. Leifsson, S. and Yıldırım, Z., At the Crossroads of Biology and Electronics: Exploring the Transformative Impact of Bioelectronic Systems on Medical,

Environmental, and Security Frontiers. *Int. J. Responsible Artif. Intell.*, *14*, 3, 2024.

4. Tohirov, U., Artificial Intelligence in Healthcare. *Sci. Soc.*, *1*, 2, 54–62, 2023.
5. Omondi, O.B., Arroyan, Y.N., Onyango, B., Kong, L., Wang, G., Ye, Z., Revolutionizing healthcare: Emerging frontiers in 3D bioprinting of tissues and organs. *Eur. Polym. J.*, 113210, 2024.
6. Ng, J.Y., Cramer, H., Lee, M.S., Moher, D., Traditional, complementary, and integrative medicine and artificial intelligence: Novel opportunities in healthcare. *Intgr. Med. Res.*, 101024, 2024.
7. Majrashi, M.A.A., Yahya, E.B., Mushtaq, R.Y., HPS, A.K., Rizg, W.Y., Alissa, M., Hosny, K.M., Revolutionizing Drug Delivery: Exploring the Impact of Advanced 3D Printing Technologies on Polymer-Based Systems. *J. Drug Deliv. Sci. Technol.*, 105839, 2024.
8. Abbas, S. and Costa, L., Biomedical Engineering: Bridging the Gap between Technology And Medicine. *Res. Med. Sci. Rev.*, *1*, 02, 128–136, 2023.
9. Herrington, V. and Adedeji, O., At the intersection of medical robotic surgery and drug discovery with quantum computing. *Authorea Preprints.*, 2, 3, 274–281, 2023.
10. Abroro'g'li, L.Y., The Significance of Advanced Technological Processes in Tailoring Stents for Optimal Biocompatibility. *World Sci. Res. J.*, *19*, 1, 69–71, 2023.
11. Adeleke, A.K., Olu-lawal, K.A., Montero, D.J.P., Olajiga, O.K., Ani, E.C., The intersection of mechatronics and precision engineering: Synergies and future directions. *Int. J. Sci. Res. Arch.*, *11*, 1, 2356–2364, 2024.
12. Singam, A., Revolutionizing patient care: A Comprehensive review of artificial intelligence applications in anesthesia. *Cureus*, *15*, 12, 2023.
13. Emeihe, E.V., Nwankwo, E., II, Ajegbile, M.D., Olaboye, J.A., Maha, C.C., Revolutionizing drug delivery systems: Nanotechnology-based approaches for targeted therapy. *Int. J. Life Sci. Res. Arch.*, *7*, 1, 040–058, 2024.
14. Zhang, S., Kong, N., Wang, Z., Zhang, Y., Ni, C., Li, L., Yan, F., Nanochemistry of gold: from surface engineering to dental healthcare applications. *Chem. Soc. Rev.*, 2024.
15. Singh, B. and Kaunert, C., Integration of Cutting-Edge Technologies such as Internet of Things (IoT) and 5G in Health Monitoring Systems: A Comprehensive Legal Analysis and Futuristic Outcomes. *GLS Law J.*, *6*, 1, 13–20, 2024.
16. Tariq, M.U., Advanced Wearable Medical Devices and their Role in Transformative Remote Health Monitoring, in: *Transformative Approaches to Patient Literacy and Healthcare Innovation*, pp. 308–326, IGI Global, 2024. https://www.igi-global.com/chapter/advanced-wearable-medical-devices-and-their-role-in-transformative-remote-health-monitoring/342833.
17. Hatem, N.A., Pharmacointelligence: Revolutionizing pharmacy practice in the Intelligence-driven healthcare Era. Available SSRN, 4754711, 2024. http://dx.doi.org/10.2139/ssrn.4754711.

18. Milson, S. and Derrick, R., *Microfluidics and Nanotechnology: a Synergistic Approach for Bioengineering (No. 11271)*, EasyChair, 2023. https://easychair.org/publications/preprint/VwFP.
19. Afantitis, A. and Lynch, I., Computational and structural biotechnology meets nanoscience and advanced materials. *Comput. Struct. Biotechnol. J.*, *25*, 1–2, 2024.
20. Haq, A. and Karim, A., Analyzing Market Trends with Advanced Data Analytics in Pure Science. *Open Horizon Sci. Rev.*, *1*, 01, 11–20, 2023.
21. McCourt, C., Exploring the intersection of the medical metaverse and healthcare ethics: future considerations and caveats. *Global Health J.*, 8, 1, 36–40, 2024.
22. D'Urso, F., Piscitelli, P., Miani, A., Broccolo, F., Revolutionizing Medical Education: The Impact of Virtual and Mixed Reality on Training and Skill Acquisition, 2023.
23. Akindote, O.J., Adegbite, A.O., Dawodu, S.O., Omotosho, A., Anyanwu, A., Maduka, C.P., Comparative review of big data analytics and GIS in healthcare decision-making. *World J. Adv. Res. Rev.*, *20*, 3, 1293–1302, 2023.
24. Tula, O.A., Daraojimba, C., Eyo-Udo, N.L., Egbokhaebho, B.A., Ofonagoro, K.A., Ogunjobi, O.A., Banso, A.A., Analyzing global evolution of materials research funding and its influence on innovation landscape: a case study of us investment strategies. *Eng. Sci. Technol. J.*, *4*, 3, 120–139, 2023.
25. Zaki, H., Shayea, I., Saoud, B., Azmi, M.H., El-saleh, A.A., Saad, S.A., Alankhi, M., Advancing Patient Outcomes: the Intersection of Biomedical Engineering, Medical Devices, and AI. *EasyChair Prepr.*, 31, 9, 12174, 2024.
26. Alshangiti, D.M., El-Damhougy, T.K., Zaher, A., Madani, M., Revolutionizing biomedicine: advancements, applications, and prospects of nanocomposite macromolecular carbohydrate-based hydrogel biomaterials: a review. *RSC Adv.*, *13*, 50, 35251–35291, 2023.
27. Ighodaro, A., Osarobo, J.A., Onuguh, I.C., Ogbeide, O.K., Ifijen, I.H., Challenges and Future Perspectives of Biomimetic Materials for Biomedical Applications: Bridging the Gap Between Nature and Medicine, in: *TMS Annual Meeting & Exhibition*, pp. 877–896, Springer Nature Switzerland, Cham, 2024, February.
28. Dhoundiyal, S., Srivastava, S., Kumar, S., Singh, G., Ashique, S., Pal, R., Taghizadeh-Hesary, F., Radiopharmaceuticals: navigating the frontier of precision medicine and therapeutic innovation. *Eur. J. Med. Res.*, *29*, 1, 26, 2024.

4

Nanoscopic Marvels: Exploring Carbon Nanoparticles in Biomedicine

Ferlin Deva Shahila* **and Ashwini A.**

Electronics and Communication Engineering, Vel Tech Rangarajan Dr. Sagunthala R&D Institute of Science and Technology, Chennai, Tamil Nadu, India

Abstract

The state of knowledge on the present and potential uses of carbon Nanoparticles in medicine is incessantly evolving, with ongoing research exploring their applications in drug delivery, diagnostics, and therapeutic treatments. Numerous members of the carbon nanoparticle group have distinct physicochemical characteristics that make them desirable aspirants for utilization throughout the realm of clinical medicine. Graphene, graphene oxide, and carbon nanotubes are the most well-known (and studied) carbon nanoparticles. The primary application areas are molecular imaging and the identification of substances or coumarin found within the body through medical diagnostics. The utilization of nanoparticles in regenerative medicine and therapies, such as drug delivery, tissue engineering, and cell scaffolding, as well as the formulation of antibodies, remains under investigation. Numerous *in vivo* investigations are currently underway because the issue of nanoscale toxicity is still open. The conclusions indicate that nanoparticles that have been deployed do not have substantial harmful consequences; nonetheless, extreme vigilance should be required prior to introducing microscopic particles into ordinary clinical roles. These kinds of studies reveal that ceramic nanoparticles feature enormous promise for both medical diagnosis and care. In technologies and science, small particles inspired by carbon, or carbon-based nanoparticles (CBNs), are becoming more and more desirable. Materials based on carbon have recently gained value. Many distinct kinds of carbon exist, extending from the well-known allotropic phases of amorphous carbon, graphite, and diamonds to the just discovered auspicious fullerene, graphene oxide (GO), diamond quantum dots (GQDs), and carbon nanotubes (CNTs). Because of their unique

**Corresponding author*: franklinferlin@gmail.com

Kolla Bhanu Prakash, S. V. Ranganayakulu and K. S. Jagannatha Rao (eds.) New Frontiers in Materials Science: Interdisciplinary Approaches to Innovation and Technologies, (61–76) © 2025 Scrivener Publishing LLC

characteristics, all members of the carbon family have found extensive use in a variety of biological applications, including the administration of medicines, tissue engineering, imaging, biological detection testing, and cancer treatment.

Keywords: Carbon nanoparticles, graphene, MWCNT, biomedical applications, diagnostic, therapy

4.1 Introduction

Arguably, the most abundant carbon appears in minerals on Earth. It comes in numerous larger compounds, such as proteins, carbohydrates, and DNA, among others, that are important to life. Carbon exists in various allotropes and pure forms. Undefined graphite, Lampblack, coal, and charcoal are examples of such alternate forms. Crystalline carbon allotropes that occur naturally include gems as well as graphite. Carbon-based nanoparticles capitalize on the unique features of carbon. During the 1980s, Carbon Nanoparticles (CNPs) first emerged [1, 2]. Different types of carbon are represented by CNPs, such as nanodiamonds, fullerenes, carbon nanotubes, carbon nanohorns, graphene, and diamond quantum dots, as well as arbitrary nanoparticles composed of carbon, ultrafine carbon particles, as well as the dots of carbon [3].

CNPs have extraordinary mechanical properties (such as extreme stiffness, strength, and toughness), as well as exceptional electrical and heat conductance, stability, and minimal toxicity because they are comprised entirely of carbon. Furthermore, CNPs are exceptionally hydrophobic due to their sp2 hybridization [4]. The skeletal parent of graphene includes many different carbon corundum such as nano-ring carbon, graphite, single, double, and multi-walled tunnel fibers, and graphene, making it the category of CNPs with the most significant research.

CNPs have been thoroughly investigated, taken outside of labs, and produced on a large scale. Figure 4.1 shows the main types of carbon nanoparticles. Different techniques are used in the commercial manufacturing of micron-sized particles, including tunnels and diamonds. and those constituted by technical cleavage from natural graphite, chemical evaporation deposition, chemical exfoliation of raw graphite, including plasma, or the epidermal growth of Graphene is the most common way for creating graph in significant numbers on the inner face of silicon carbide [5]. Reduction Graphene Oxide (RGO) and Graphene Oxide (GO) are the two most widely utilized graphene derivatives, and they are made by oxidizing graphene.

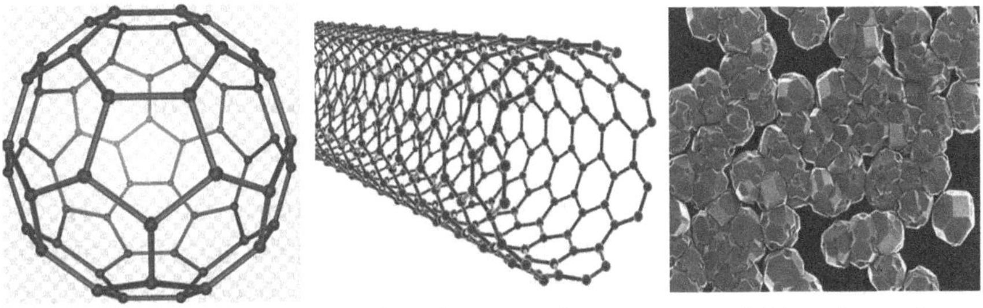

(a) Fullerenes (b) Carbon Nanotubes (c) Nano diamonds

Figure 4.1 Main types of carbon nanoparticles.

These procedures involve treating graphene with oxidizing chemicals, such as KMnO4, and protonated solvents such as phosphoric, sulfuric, or nitric acid, or their combinations. This process causes a covalent bond that links the moieties with carboxyl and methyl to form graphene. After that, a reduction is performed to create RGO, during which the majority of the functional groups involving oxygen are eliminated using a variety of techniques, such as treating the compound with hydrazine hydrate or vitamin C [6, 7].

Stretching graphene sheets yields carbon nanotubes (CNTs), which are nanomaterials with cylindrical tubular structures and a nanometer-diameter. There are two kinds of CNTs in accordance with the total amount of layers in the graphite sheets: single-walled and multi-walled [8]. Arc discharge process, pulsed ablation with lasers, and chemical vacuum deposition (CVD) are the most ubiquitous processes used to create carbon nanotubes. Carbon-containing precursor gases, such as CO_2 or C_2H_2, C_2H_4, and other hydrocarbons, are used in the CVD process. During CVD, the temperature ranges between 350 and 1000 °C. The formation of CNTs can be influenced by multiple factors, such as temperature, reaction duration, catalyst's dimension of particles, and type and the impact of the gas's velocity [9].

A set of Graphite rods are utilized as the electrodes in the arc discharge process, connecting to the cathode and anode. After that, an inert gas (helium) is poured over the electrodes to apply the direct current arc voltage. CNTs and fullers are produced using this process. For example, multi-walled carbon nanotubes are produced on the cathode, and soot featuring fullerenes is left throughout the chamber [10]. Graphite ablation by pulsed laser is a very promising technique. It generates a highly natural product with the required size and shape quickly and cheaply, all without the need for a catalyst or vacuum.

4.2 Applications

4.2.1 Biological Imaging, Pathology-Related Detection, and Diagnostics

Correct prompt diagnosis—whether for tumors or tissue damage—is critical to the success of the treatment. CNPs might significantly improve the average level of medical imaging. When photoexcited, the majority of CNPs naturally produce fluorescence [11, 12].

There is widespread use of carbon nanotubes, carbon dots, and graphene quantum dots. Most importantly, their incredibly vast surface area helps contain numerous molecules and provide multiplex sensing. In contrast, CNPs (carbon dots smaller than 10 nm) that rapidly excrete through the renal or hepatobiliary tract are appropriate for biological imaging [13]. The luminescence may be focused, and individual organelles and cells photographed pursuant to the distinct CNP customization and nanocrystals.

Customized carbon dots are injected intravenously into mice, where they quickly penetrate the conglomeration and concentrate inside the kidney, liver, and lung, in addition to breaching the blood-brain barrier [14]. Single-wall carbon nanotube-coated M13 virus has the ability to specifically target tumor cells. Additionally, PET, or positron emission tomography, can detect Technetium-99 m Tc radionuclide-labeled particles that target tumors [15, 16]. Through these techniques, imaging for healthcare technologies such as MRI, which is magnetic resonance imaging, and computed tomography, or CT for short, and PET can be combined with the fluorescent effect of CNPs. This combination improves contrast and displays more details with higher quality and sensitivity in imaging [17]. The effective use of CNPs during surgery is widely established. During surgical gastrectomy with lymph node dissection, Li *et al.* investigated the transit noteworthy buildup of CNPs (mean 150 nm) in the lymph nodes of patients suffering from aggressive gastrointestinal cancer. To lower the chance of a cancer recurrence, metastatic lymph node identification and staining is crucial. Thirty patients took part in the research. Following the administration of intravenous fluid, no significant adverse effects were reported. Black coloration was caused by CNPs that collected specifically in the draining nodes of the lymph system. Because fewer lymph nodes could be detected by this approach, there were more lymph nodes that could be harmful [18]. Postoperative identification of lymph nodes for individuals with colorectal, breast, and thyroid cancers has been shown to have the same impact [19–21]. CNPs aid in the differentiation of parathyroid glands and lymph nodes in thyroid carcinoma [22]. Crucially, before and during surgery, CNPs that were utilized in colon cancer treatment robotic tattooing enable the exact location and identification of recurrent neoplastic lesions. This lessens surgical trauma by making it possible to carry out an exact resection [23]. Since the CNP transistor and the CNP-featured content have excellent electrical permeability and great cardiac electrical signal amplification, they can be utilized in ventricular electrophysiological signal recording electrodes.

MWCNTs, which are multi-walled carbon nanotubes, or graphene are frequently employed in hospitals for both long-term and short-term monitoring. These devices are wearable and track changes in an electroencephalogram (ECG) in real-time as people go about their daily lives [24, 25].

4.2.2 Diagnostics Photoacoustic

"The Application of Nanotechnology for the Diagnosis and Treatment of Endocrine Disorders" looks into how endocrine disorders including diabetes, thyroid issues, and other diseases linked to hormones can be better diagnosed and treated with the use of nanotechnology. Figure 4.2 shows diagrammatic representation of the photoacoustic imaging concept. It focuses on how tailored medication delivery systems might be enhanced by nanoparticles, increasing the non-intrusive biomedical imaging technique known as photoacoustic imaging (PA). A may be used to diagnose and treat a variety of cancers, such as breast, cutaneous, and colorectal cancer [26]. PA also includes photoacoustic microscopy and photoacoustic computed tomography. With its excellent temporal and spatial resolution and relatively deep tissue imaging capacity, PA is useful for both the diagnosis and treatment of a wide range of carcinomas, including colorectal, cutaneous, and breast cancers. Specifically, when graphene oxide, graphene nanotubes, or nanodots are paired with natural polymers, metal

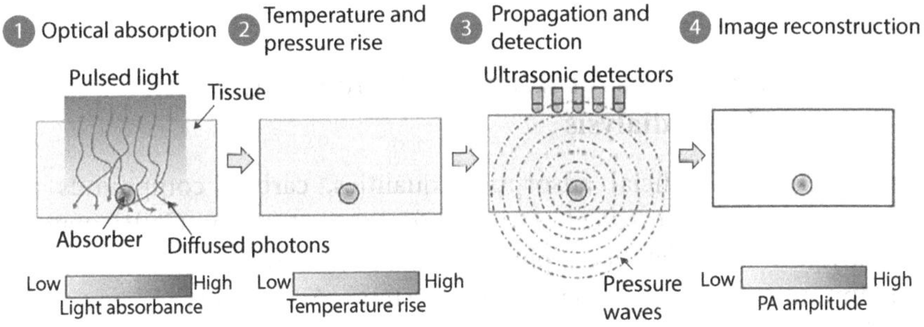

Figure 4.2 Diagrammatic representation of the photoacoustic imaging concept.

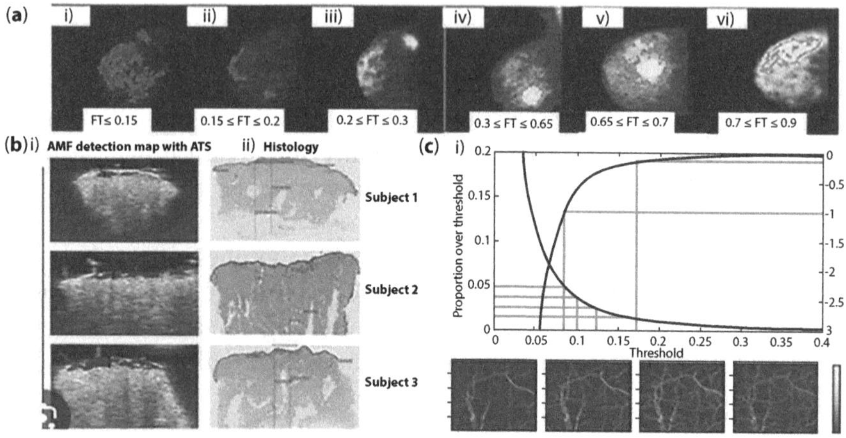

Figure 4.3 Schematic diagram of photoacoustic imaging.

nanoparticles, or pigments, they make effective contrasting agents [27]. Certain tissues can be targeted by these contrast agents; different types of tumors specifically absorb Arg-Gly-Asp cyclic peptides on single-walled carbon nanotubes [28]. *In vitro* and *in vivo*, Han *et al.* reported ultra-small particles of carbon made of nearly infrared II photothermal material that exhibit multicolored fluorescence and generate photoacoustic signals [29]. These dots can easily penetrate tumor tissue.

Figure 4.3 shows the schematic diagram of photoacoustic imaging. The principle of photoacoustic imaging schematic diagram: (1) Tissues absorb light energy when exposed to pulsed light at a specific angle; (2) The absorbed light energy becomes converted into thermal energy, momentarily raising the level of humidity before thermoelastic expansion causes the surrounding pressure to rise; (4) From the spot that is of interest, photoacoustic waves delivering pressure spread towards the exterior, which are detected by ultrasonic detectors; and (5) pictures are produced by reconstructing the propagating signals.

4.2.3 Hem Sorbents for Adsorption, Hemofiltration, and Hemodialysis

Due to their beneficial adsorption qualities, carbon compounds are frequently used in medical settings. To lessen toxicity, CNPs may be employed to assimilate ions, pharmaceuticals, toxins, metabolites, lipids, and immune proteins from circulatory systems. According to Zheng *et al.*'s work, cytokines that contribute to inflammation (TNF-α, IL-6, and IL-8) and can lead to sepsis and multiorgan failure can be adsorbed by graphene nanoplatelets. The outcome of hemoperfusion may be enhanced by their inclusion into a flexible polytetrafluoroethylene (PTFE) film. Figure 4.4 shows the schematic diagram of hemofiltration. Proinflammatory cytokines were confirmed to be absorbed by graphene-based nanomaterials by Berdych *et al.* Granulated graphene nanoplates and nanoplatelets containing PTFE were employed. High removal efficiencies—up to 100% in some situations—are possessed by both kinds of CNPs. Greater uptake capabilities for cytokines that foster disease are also present in nanodiamonds. According to Yoo *et al.*'s research, inflammatory cytokines are substances including TNF-α, IL-1β, IL-4, IL-6, and IFN-γ can be absorbed by nanodiamonds, which reduces inflammation and lowers the risk of sepsis-related multiorgan failure and mortality. Wu *et al.* created chitin/carbon nanotube microspheres immobilized with lysine and tested their ability to extract human blood containing hemoglobin. The amount of hemoglobin in the plasma indicates hemolysis or hepatobiliary system malfunction. Bilirubin encephalopathy is brought on by bilirubin's toxicity to the brain. Situated bile adsorption capability and enhanced bilirubin clearance were demonstrated by the produced CNPs. An *et al.* took advantage of CNPs' exceptional blood-brain barrier absorption efficiency. They used polyether sulfone and multi-walled carbon nanotubes featuring carboxyl functionalization (MWCNT) to create a hemodialysis membrane. Excellent blood compatibility was demonstrated by this membrane; urea and creatinine clearance were greatly enhanced, and membrane fouling was reduced by 30%. Additionally, cholesterol is a risk factor for a number of disorders.

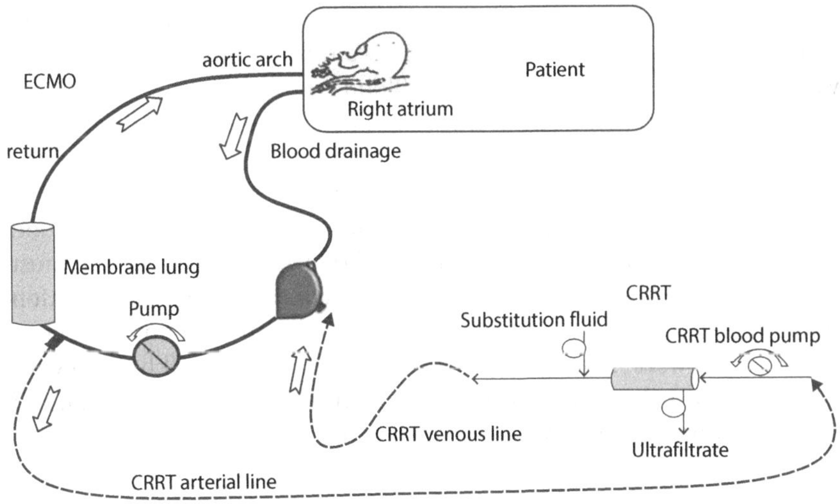

Figure 4.4 Schematic diagram of hemofiltration.

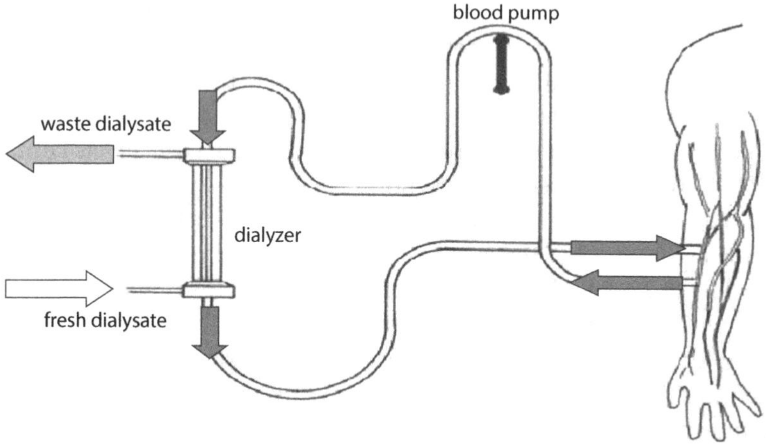

Figure 4.5 Schematic diagram of hemodialysis.

Figure 4.5 shows the schematic diagram of hemodialysis. Your blood goes through an external filter known as a dialyzer during hemodialysis. One term used when referring to a dialyzer is "artificial kidney". There are two needles inserted by a dialysis nurse into your arm or professional to start obtaining hemodialysis. Under the instruction of your medical staff, you could decide to insert the needles yourself. If needle insertion hurts, an analgesic cream or spray is an option. Every needle has a soft tube attached to it that is connected to the dialysis unit. The pumped blood passes through the gadget's filter and then returns to your body after dialysis. Your cardiovascular system is monitored by the dialysis machine, and determines the values. As the blood passes through the filter, the body's waste and excess fluids are eliminated. Blood is pushed through numerous, incredibly thin, hollow fibers as it goes through one end of the filtration. In the exact opposite direction from the blood that flows on the surface, the dialysis solution moves through the hollow fibers' exterior. Contamination products from your blood permeate the dialysis solution. Filtered blood then goes back into your body through the hollow fibers. For therapy, most people visit a dialysis facility. The medical staff at the dialysis institution set up and support you in connecting to the dialysis equipment. A crew of healthcare workers will be on hand to assist you. You will head to your physician as usual. A social worker, a dietitian, a healthcare technician, and nurses are possible additional team members.

4.2.4 Photodynamic Therapy

A medication called a photosensitizer is used in conjunction with light energy in a two-stage treatment called photodynamic therapy (PDT), which is intended to kill malignant and precursory tumor cells upon light activation. Materials sensitive to sunlight are triggered by a particular light spectrum, often emitted by a light source. The photo detector is safe until light forces it to become active. On the other hand, the photosensitizer becomes poisonous to the targeted tissue upon light activation.

Today, there are several photosensitizer medications available to treat quite a few conditions under certain circumstances, such as age-related macular degeneration, psoriasis, acne, and numerous malignancies of the skin, lungs, esophagus, bile duct, spleen, brain, bowels, and head and neck. PDT assists in the medical management of bacterial, fungal, and viral infections in addition to these ailments. Studies indicate that the technique is based on light can activate the defense system, providing the body's immunological system with an additional tool to help eliminate precancerous and cancerous cells.

The physician applies a cream containing a light-sensitive agent to the skin carcinoma and its surrounding tissue. Figure 4.6 shows the schematic diagram of photodynamic therapy. The chemical may also be supplied as a pill or injection. After that, your doctor treats the area in question with a specific light that kills cancer cells. A highly popular and well researched photosensitizer is profiler sodium, often known as photoprint. Red light from a laser triggers it. It is being studied for use in various cancer types and has been approved by the Food and Drug Administration (FDA) for the therapeutic use of individuals with specific forms of lung and esophageal tumors.

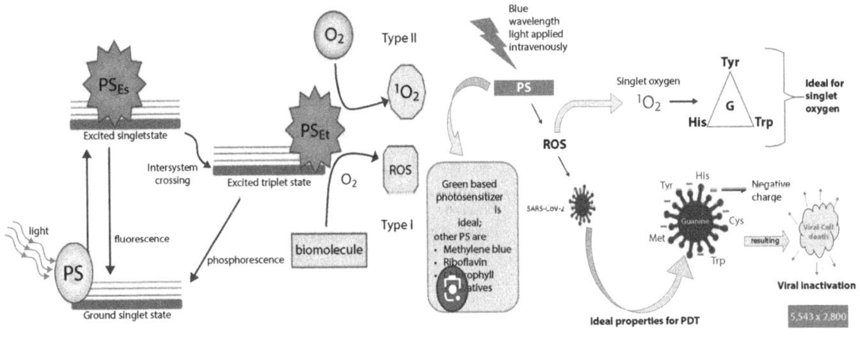

Figure 4.6 Schematic diagram of photodynamic therapy.

Using light-sensitive medications called photosensitizers in conjunction with particular light wavelengths, photodynamic therapy (PDT) is a minimally invasive treatment that kills malignant or aberrant cells. Following administration, the photosensitizer builds up in the target tissue and releases reactive oxygen species that harm and ultimately disable cells when exposed to a certain light source. PDT is used to treat non-cancerous illnesses such as macular degeneration, as well as a variety of malignancy, such as those of the skin, lungs, and esophagus. Compared to conventional treatments like chemotherapy or radiation, it has fewer side effects, less harm to nearby healthy cells, and more specific targeting of the afflicted area. Type I and Type II responses during photodynamic treatment are illustrated in Jablonski's schematic graphic illustrates the mechanism of action of PDT. After light is absorbed, the photosensitizer (PS) reaches an excited singlet state. Following intersystem crossing, the PS, which is presently stimulated to the triplet state, has two possible reactions: either directly with oxygen by transporting energy, producing singlet oxygen on its own (Type II reaction), or directly sharing hydrogen atoms (electrons) with biomolecules, forming reactions that produce ROS (type I reaction). When they participate with molecular oxygen, photosensitizer (PS), reactive oxygen species (ROS), singlet oxygen (1 O2), PS excited triplet state (PSET), and PS excited singlet state (PSEs) are produced.

To inactivate the coronavirus that causes SARS-CoV-2, the severely catastrophic respiratory illness, photodynamic therapy (PDT) may prove to be an effective treatment for COVID-19. The COVID-19 pandemic, which was brought on by the spread of the SARS-CoV-2 virus, has claimed many lives globally because there are no viable therapies. Although the FDA has recently approved two mRNA-based vaccinations for use in emergencies, there is still a shortage of vaccines. Finding novel therapeutic approaches is therefore essential for the care of COVID-19 patients. PDT has been used for many years to deactivate germs, and it may be a useful and promising treatment for COVID-19 patients. A photosensitizer, light, and oxygen make up the PDT process, which produces a localized burst of reactive oxygen.

This therapy can enhance your beauty while also safeguarding your health. Your skin could have a smooth, appealing texture after your precancerous growths are removed. Furthermore, this treatment might help with pigmentation problems, rosacea, sun damage, and other cosmetic concerns.

4.2.5 Vaccine Production

For a vaccine to stop the disease from spreading or, at the very least, from progressing severely, the chosen antigen must provoke an immune reaction against it. Vaccines are used increasingly frequently these days not only to treat autoimmune illnesses and cancer but also to prevent infections. As antigen carriers and adjuvants, CNPs may be laden with tumor or bacterial target antigens as well as immunological features that boost the development memory T and B cells, antibody generation, and the immune system's reaction.

Figure 4.7 shows the schematic diagram of vaccine production. Carbon nanotube-binding antigens can enhance T lymphocyte cytolysis of malignant cells or cause the generation of certain IgGs that neutralize a pathogen in animal models. The creation of the graphene oxide (GO) influenza vaccine nanoplatform functioning with polyethyleneimine, which bonds to influenza heterologous hemagglutinin, was detailed in an intriguing study by Dong *et al.* Intranasally administered vaccinations prevented infection in mice following the viral challenge by eliciting an extraordinary humoral and cellular immune response. Additionally, nanodiamonds can act as a carrier of influenza viral antigens. They can triple the synthesis of neutralizing antibodies in mice.

A wide range of concepts, from sophisticated molecules generated by molecular biology to stunted creatures, are used in the manufacturing of vaccines. It is true that the most complex processes are not always better, just as the most basic techniques rarely produce meaningful products.

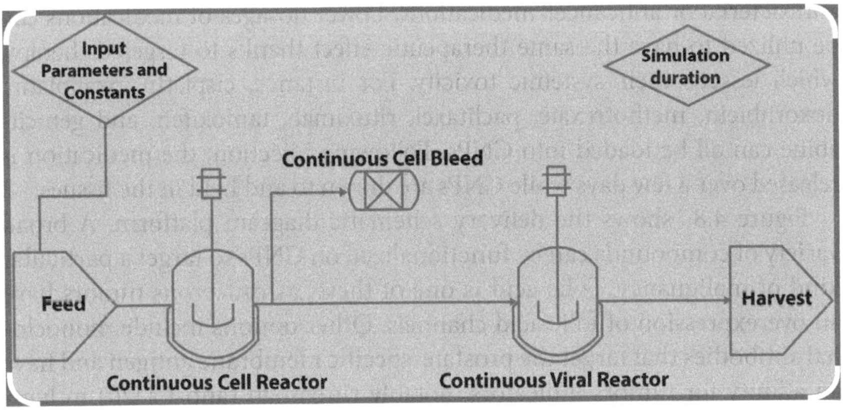

Figure 4.7 Schematic diagram of vaccine production.

4.2.6 Therapy—Platform of Delivery and Intervention

Drug therapy is institutionalized, meaning it affects the entire body. Significant amounts of substances that cause violence against the body are frequently required to get the suitable proportion at the effect's location (like a tumor site). Targeted therapy has the potential to significantly improve efficacy and lower systemic inflammation. For the transport of drugs to different tissues, CNPs are (almost) the ideal carrier. They possess large loading capabilities for various pharmaceutical compounds, including mixtures of them. The ability to target particular organs or molecules, improve penetration and retention impacts, and regulate drug release are all made possible by the modification of CNPs. Temperature, light, pH, and enzymatic activity are examples of release stimuli.

The process of introducing therapeutic genes or medication components into host cells to produce a therapeutic effect is known as gene therapy. Selecting the best vector to deliver specific amounts of foreign gene products to each target cell in a way that permits expression without causing cytotoxicity is our aim. Cellular barriers—such as cellular absorption, endosomal escape, nuclear entrance, and nucleic release—and extracellular barriers—such as particle clearance processes, nucleic acid breakdown, and specialized targeting—must be overcome by the ideal gene delivery vector. In accordance with good manufacturing practices (GMP) guidelines, Creative Biolabs offers a comprehensive selection of gene therapy platforms to produce gene delivery vectors that are efficient, targeted, safe, and long-lasting for use by researchers throughout the globe in the treatment of illnesses. Carbon nanotubes with functionalities, such as graphene, fullerene, and others, have the ability to bind proteins (such as monoclonal and polyclonal antibodies) or antibacterial or anticancer medications. Lower dosages of medications can be utilized to have the same therapeutic effect thanks to targeted therapy, which lessens their systemic toxicity. For instance, cisplatin, oxaliplatin, doxorubicin, methotrexate, paclitaxel, rituximab, tamoxifen, and gemcitabine can all be loaded into CNPs. Following injection, the medication is released over a few days while CNPs are drawn to and held in the tissues.

Figure 4.8 shows the delivery schematic diagram platform. A broad variety of compounds can be functionalized on CNPs to target a particular kind of malignancy. Folic acid is one of these, as cancerous tumors have an overexpression of folic acid channels. Other options include monoclonal antibodies that target the prostate-specific membrane antigen and have an affinity for tumors indicators, notably rituximab (anti-CD20) in lymphomas or anti-HER2 in breast cancer. To treat oral squamous cell carcinoma, nanoscale GOs partnered with HN-1 and doxorubicin (a particular

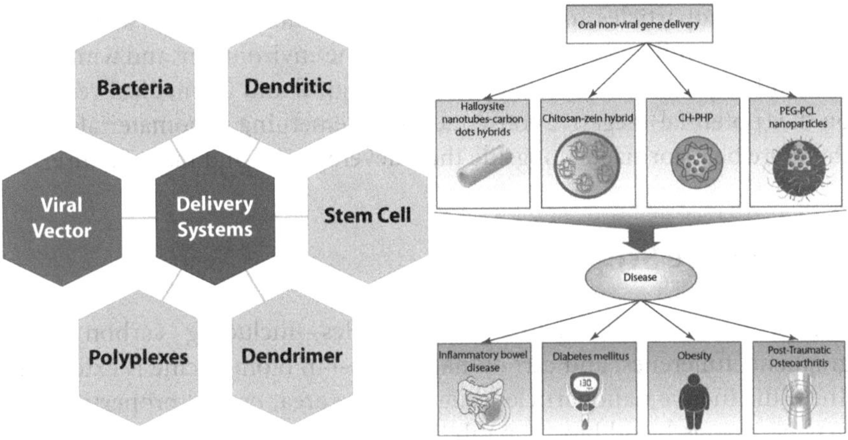

Figure 4.8 Delivery schematic diagram platform.

antigen) were successful. When compared to free doxorubicin, the particles' toxicity was much higher due to their easy absorption by cancer cells. Bromocriptine or doxorubicin-loaded MWCNTs can cause lung and prostate cancer cells to undergo necrosis or apoptosis. These effects can last for several days.

In addition to anticancer medications, CNPs can also transport antimicrobial substances (antibiotics, antifungals, antiparasitic, and antiviral medications) and growth factors, such as vascular endothelial growth factor (VEGF), to promote regeneration in ischemic heart tissue. CNPs can eradicate viral infections, particularly bacterial or fungal ones, and circumvent antibiotic-resistant bacteria, such as *Salmonella typhimurium*, *Escherichia coli*, *Staphylococcus aureus*, *Streptococcus mutans*, *Pseudomonas aeruginosa*, *Klebsiella pneumoniae*, and *Porphyromonas gingivalis*, even in the absence of antimicrobial drug loading. The creation of Ag nanocomposites strengthens their antibacterial properties.

4.3 Conclusions

We only covered the most significant carbon nanoparticles and their potential medical uses in the review that was provided. Due to their numerous applications, efficacy, and affordability compared to their performance, carbon nanoparticles might find widespread use in medicine. The usage of CNPs raises questions regarding their biosafety, despite the fact that their benefits are well-established.

Since nanoparticles are a relatively recent phenomenon, not all the variables that can influence how they behave in the environment and within the human body are now understood. Toxicological and biomedical research on the (potential) negative consequences of emerging nanomaterials must therefore be prioritized alongside their development and manufacture.

4.4 Future Scope

The scope of carbon-based nanoparticles—including carbon dots, graphene, fullerenes, and carbon nanotubes—in biomedicine is extensive. These distinctive nanoparticles' huge surface area, optical properties that may be changed, and biocompatibility make them crucial for tissue engineering, drug delivery, bioimaging, and biosensing. While the study highlights their potential for use in cancer treatment, targeted drug delivery, and disease monitoring, it also highlights problems that need to be fixed to fully utilize them in medical applications, such as toxicity, stability, and scalability. The use of carbon-based nanoparticles, such as carbon dots, fullerenes, graphene, and carbon nanotubes, in biomedicine is examined in the study "Nanoscopic Marvels: Exploring Carbon Nanoparticles in Biomedicine". These unique nanoparticles are useful for drug administration, bioimaging, biosensing, and tissue engineering because of their large surface area, modifiable optical characteristics, and biocompatibility.

References

1. Sharma, M., Understanding the mechanism of toxicity of carbon nanoparticles in humans in the new millennium: A systemic review. *Indian J. Occup. Environ. Med.*, 14, 3–5, 2010, [CrossRef] [PubMed].
2. Kroto, H.W., Heath, J.R., Brien, S.C.O., Curl, R.F., Smalley, R.E., C60: Buckminsterfullerene. *Nature*, 318, 162–163, 1985, [CrossRef].
3. Chen, C. and Haifang, W., *Biomedical Applications and Toxicology of Carbon Nanomaterials*, Wiley, Hoboken, NJ, USA, 2016, [CrossRef].
4. Choudhary, N., Hwang, S., Choi, W., Carbon Nanomaterials: A Review, in: *Handbook of Nanomaterials Properties*, pp. 709–769, Springer, Berlin/Heidelberg, Germany, 2014, [CrossRef].
5. Lin, L., Peng, H., Liu, Z., Synthesis challenges for graphene industry. *Nat. Mater.*, 18, 520–524, 2019, [CrossRef].
6. Smith, A.T., LaChance, A.M., Zeng, S., Liu, B., Sun, L., Synthesis, properties, and applications of graphene oxide/reduced graphene oxide and their nanocomposites. *Nano Mater. Sci.*, 1, 31–47, 2019, [CrossRef].

7. Habte, A.T. and Ayele, D.W., Synthesis and Characterization of Reduced Graphene Oxide (rGO) Started from Graphene Oxide (GO) Using the Tour Method with Different Parameters. *Adv. Mater. Sci. Eng.*, 2019, 5058163, 2019, [CrossRef].
8. Eatemadi, A., Daraee, H., Karimkhanloo, H., Kouhi, M., Zarghami, N., Akbarzadeh, A., Abasi, M., Hanifehpour, Y., Joo, S.W., Carbon nanotubes: Properties, synthesis, purification, and medical applications. *Nanoscale Res. Lett.*, 9, 393, 2014, [CrossRef].
9. Wang, X.-D., Vinodgopal, K., Dai, G.-P., Synthesis of Carbon Nanotubes by Catalytic Chemical Vapor Deposition, in: *Perspective of Carbon Nanotubes*, IntechOpen, London, UK, 2019, [CrossRef].
10. Arora, N. and Sharma, N., Arc discharge synthesis of carbon nanotubes: Comprehensive review. *Diam. Relat. Mater.*, 50, 135–150, 2014, [CrossRef].
11. Ismail, R.A., Mohsin, M.H., Ali, A.K., Hassoon, K.I., Erten-Ela, S., Preparation and characterization of carbon nanotubes by pulsed laser ablation in water for optoelectronic application. *Phys. E Low Dimens. Syst. Nanostruct.*, 119, 113997, 2020, [CrossRef].
12. Liu, F., Gao, Y., Li, H., Sun, S., Interaction of propidium iodide with graphene oxide and its application for live cell staining. *Carbon*, 71, 190–195, 2014, [CrossRef].
13. Liu, J., Li, D., Zhang, K., Yang, M., Sun, H., Yang, B., One-Step Hydrothermal Synthesis of Nitrogen-Doped Conjugated Carbonized Polymer Dots with 31% Efficient Red Emission for *In Vivo* Imaging. *Small*, 14, e1703919, 2018, [CrossRef].
14. Liu, J., Li, R., Yang, B., Carbon Dots: A New Type of Carbon-Based Nanomaterial with Wide Applications. *ACS Cent. Sci.*, 6, 2179–2195, 2020, [CrossRef].
15. Patel, K.D., Singh, R.K., Kim, H.-W., Carbon-based nanomaterials as an emerging platform for theranostics. *Mater. Horiz.*, 6, 434–469, 2019, [CrossRef].
16. Ghosh, D., Bagley, A.F., Na, Y.J., Birrer, M.J., Bhatia, S.N., Belcher, A.M., Deep, noninvasive imaging and surgical guidance of submillimeter tumors using targeted M13-stabilized single-walled carbon nanotubes. *Proc. Natl. Acad. Sci. U.S.A.*, 111, 13948–13953, 2014, [CrossRef].
17. Cherukula, K., Manickavasagam Lekshmi, K., Uthaman, S., Cho, K., Cho, C.-S., Park, I.-K., Multifunctional Inorganic Nanoparti- cles: Recent Progress in Thermal Therapy and Imaging. *Nanomaterials*, 6, 76, 2016, [CrossRef].
18. Li, Z., Zhaode, B., Bu, Z., Wu, A., Wu, X., Shan, F., Ji, X., Zhang, Y., Xing, Z., Ji, J., Clinical study of harvesting lymph nodes with carbon nanoparticles in advanced gastric cancer: A prospective randomized trial. *World J. Surg. Oncol.*, 14, 88, 2016, 17 of 21.

19. Ma, J.-J., Zhang, D.-B., Zhang, W.-F., Wang, X., Application of Nanocarbon in Breast Approach Endoscopic Thyroidectomy Thyroid Cancer Surgery. *J. Laparoendosc. Adv. Surg. Tech.*, 30, 547–552, 2020, [CrossRef].
20. Liu, P., Tan, J., Tan, Q., Xu, L., He, T., Lv, Q., Application of Carbon Nanoparticles in Tracing Lymph Nodes and Locating Tumors in Colorectal Cancer: A Concise Review. *Int. J. Nanomed.*, 15, 9671–9681, 2020, [CrossRef].
21. Du, J., Zhang, Y., Ming, J., Liu, J., Zhong, L., Liang, Q., Fan, L., Jiang, J., Evaluation of the tracing effect of carbon nanoparticle andcarbon nanoparticle-epirubicin suspension in axillary lymph node dissection for breast cancer treatment. *World J. Surg. Oncol.*, 14, 164, 2016, [CrossRef].
22. Xu, S., Li, Z., Xu, M., Peng, H., The role of carbon nanoparticle in lymph node detection and parathyroid gland protection during thyroidectomy for non-anaplastic thyroid carcinoma—a meta-analysis. *PLOS ONE*, 15, 11, e0223627, 2020. https://doi.org/10.1371/journal.pone.0223627.
23. Wang, R., Zhan, H.L., Li, D.Z., Li, H.T., Yu, L., Wang, W., Application of endoscopic tattooing with carbon nanoparticlet in the treatment for advanced colorectal cancer. *Zhonghua Wei Chang Wai Ke Za Zhi*, 23, 56–64, 2020, [CrossRef] [PubMed].
24. Cui, T.-R., Li, D., Huang, X.-R., Yan, A.-Z., Dong, Y., Xu, J.-D., Guo, Y.-Z., Wang, Y., Chen, Z.-K., Shao, W.-C., *et al.*, Graphene-Based Flexible Electrode for Electrocardiogram Signal Monitoring. *Appl. Sci.*, 12, 4526, 2022, https://doi.org/10.3390/app12094526. [CrossRef].
25. Tasneem, N.T., Pullano, S.A., Critello, C.D., Fiorillo, A.S., Mahbub, I.A., Low-Power On-Chip ECG Monitoring System Based on MWCNT/PDMS Dry Electrodes. *IEEE Sens. J.*, 20, 12799–12806, 2020, [CrossRef].
26. Yan, Y., Cai, H., Yang, M., The Application of Nanotechnology for the Diagnosis and Treatment of Endocrine Disorders: A Review of Current Trends, Toxicology and Future Perspective. *Int. J. Nanomed.*, 19, 9921–9942, 2024.
27. Lipomi, D.J. and Ramji, R.S., *Introduction to Nanoengineering*, Royal Society of Chemistry, Cambridge, UK, 2024.
28. Zulfiqar, A., Ammara, F., Khan, M., Usman Ur Rehman, M., Waseem Fazal, M., Ul Wara, T., Shah, M., Akhtar, N., Two-dimensional material-based scaffolds for cell-based chip and tissue engineering and their recent progress in medical application, in: *Functionalization of Two-Dimensional Materials and Their Applications.*
29. González, P.T.P., Hernández Celi, I., Gómez Jaimes, F.N., Martínez Bonilla, C.A., Nanocellulose Composites as Chemo-/Bio-Sensing Agents and Polymeric Matrixes: A Nanotechnological Application, in: *Smart and Sustainable Applications of Nanocomposites*, pp. 192–221, IGI Global, 2024.

5

Harnessing the Power of Materials for Efficient Energy Storage and Conversion

P. Sujidha

Mohamed Sathak Engineering College, Kilakarai, India

Abstract

Recent advances in material science have significantly impacted energy storage and conversion technologies. This paper presents an overview of key trends in materials research aimed at enhancing energy storage capacities and improving energy conversion efficiencies. The discussion encompasses a variety of material types, including nanostructured materials, composites, and novel polymers, which play pivotal roles in energy storage devices such as batteries, supercapacitors, and fuel cells. Furthermore, the importance of material design at the atomic and molecular levels is emphasized, highlighting the influence of material properties on electrochemical performance. The role of advanced characterization techniques in elucidating material behavior and guiding design strategies is also explored. Additionally, emerging materials for next-generation energy technologies, such as lithium-sulfur batteries, solid-state electrolytes, and perovskite solar cells, are discussed in the context of their potential impact on the energy landscape. This review underscores the interdisciplinary nature of contemporary material science and its crucial role in addressing global energy challenges. Moreover, the utilization of nanomaterials and nanostructures has emerged as a promising approach to boost the performance of energy storage devices. Nanomaterials exhibit unique physicochemical properties that can be tailored to optimize energy storage and conversion efficiency. Additionally, the pursuit of sustainable and environmentally friendly materials has been a notable trend. Researchers are investigating materials that are abundant, low-cost, and non-toxic, aiming to reduce the environmental impact associated with energy technologies. Furthermore, advancements in material design and characterization techniques have played a pivotal role in accelerating progress in this field. Computational modeling and simulation tools have

Email: sujidhaponnaiah02@gmail.com

Kolla Bhanu Prakash, S. V. Ranganayakulu and K. S. Jagannatha Rao (eds.) New Frontiers in Materials Science: Interdisciplinary Approaches to Innovation and Technologies, (77–96) © 2025 Scrivener Publishing LLC

enabled the prediction and optimization of materials with desired properties, thus expediting the discovery process.

Keywords: Nanomaterial, energy storage, material science

5.1 Introduction to Energy Storage Materials

5.1.1 Overview of Energy Storage Systems

Energy storage systems (ESS) are indispensable assets in today's evolving energy landscape. They function as dynamic intermediaries, capable of accumulating excess energy generated during periods of low demand or high production and subsequently releasing it when required. This crucial capability is particularly valuable in integrating intermittent renewable energy sources, such as solar and wind power, into the grid. By smoothing out fluctuations in energy supply, ESSs contribute significantly to grid stability and reliability.

Furthermore, ESSs provide a versatile suite of applications beyond grid support. They can serve as backup power sources for residential, commercial, and industrial facilities, ensuring uninterrupted operations during power outages. In transportation, electric vehicles (EVs) rely on advanced battery technologies to store energy, enabling longer driving ranges and reduced emissions. Additionally, ESSs play a pivotal role in demand response programs, where they can help manage peak electricity consumption and reduce energy costs. As the world transitions toward a low-carbon future, the importance of energy storage systems is set to grow exponentially, driving innovation and development in this critical technology.

5.1.2 Importance of Material Selection

The judicious selection of materials is a cornerstone in the design and optimization of energy storage systems (ESS). These systems, crucial for the integration of renewable energy sources and grid stabilization, rely heavily on the properties of their constituent components. The performance metrics of an ESS—energy density, power density, cycle life, safety, cost, and environmental impact—are directly influenced by the materials employed.

Electrodes, the heart of energy storage, demand materials with exceptional conductivity, electrochemical activity, and stability. These properties dictate the efficiency of energy conversion and storage. Electrolytes, acting as ion conductors, must balance ionic conductivity with safety. A suitable electrolyte ensures rapid ion transport while preventing short circuits and thermal runaway. Separators, while crucial for preventing internal short circuits, should also permit efficient ion diffusion. The ideal separator strikes a balance between conductivity and porosity. Beyond the core components, the packaging material plays a pivotal role in protecting the internal components from external factors while also considering cost and environmental implications.

Material selection is a complex interplay of factors. For instance, a material with high energy density might compromise safety or cycle life. Similarly, a cost-effective material could exhibit inferior performance. Therefore, a holistic approach is essential to identify the optimal material combination for a specific ESS application.

By meticulously selecting and engineering materials, researchers and engineers can develop ESSs with superior performance, extended lifespans, and reduced environmental footprints. This advancement is instrumental in accelerating the transition to a sustainable and resilient energy future.

5.1.3 Current Trends and Innovations

The energy storage sector is undergoing a period of rapid evolution, driven by the imperative to transition to a low-carbon economy and the increasing demand for reliable, efficient, and sustainable energy solutions. Several key trends and innovations are reshaping the landscape of energy storage materials.

5.1.3.1 Transition Beyond Lithium-Ion

While lithium-ion batteries have dominated the market, limitations in terms of energy density, safety, and cost have spurred research into alternative chemistries. Solid-state batteries, with their potential for enhanced energy density, improved safety, and longer cycle life, represent a significant advancement. Sodium-ion batteries, leveraging the abundance of sodium, offer a promising avenue for large-scale stationary storage applications. Additionally, post-lithium chemistries, such as lithium-sulfur and lithium-air, are being explored for their theoretical high energy density, although significant challenges persist.

5.1.3.2 Material Optimization and Discovery

Advancements in materials science and engineering are crucial for improving energy storage performance. Sophisticated characterization techniques, including *in-situ* and operando methods, provide invaluable insights into battery mechanisms, enabling targeted material optimization. Computational modeling and simulation accelerate the discovery of novel materials with desired properties, reducing the time-consuming and expensive experimental process. Furthermore, nanomaterials and nanostructured materials offer opportunities to enhance electrochemical performance through increased surface area, improved ion transport, and faster reaction kinetics.

5.1.3.3 Sustainability and Circular Economy

The environmental impact of energy storage is a growing concern. Developing sustainable materials and processes is essential to minimize the ecological footprint of batteries throughout their lifecycle. A circular economy approach, emphasizing recycling and reuse, is vital for recovering valuable materials and reducing waste. Additionally, securing a reliable and sustainable supply chain for critical raw materials is crucial for the long-term viability of the energy storage industry.

5.1.3.4 Expanding Applications

Energy storage is playing an increasingly important role in grid modernization, transportation electrification, and portable electronics. Grid-scale energy storage systems are essential for integrating renewable energy sources, balancing supply and demand, and enhancing grid resilience. Electric vehicles rely on high-performance batteries to achieve longer driving ranges, faster charging times, and improved safety. In the realm of portable electronics, the demand for higher energy density, longer battery life, and faster charging continues to drive innovation.

5.2 Electrochemical Energy Storage

5.2.1 Battery Technologies

Battery technology stands as a cornerstone in the pursuit of a sustainable energy future. Its evolution has been instrumental in driving advancements across diverse sectors, from portable electronics to electric vehicles and grid-scale energy storage.

5.2.1.1 *Established Battery Technologies*

Lithium-Ion Batteries: This technology has undeniably dominated the market due to its impressive energy density, lightweight design, and robust cycle life. Its versatility has led to widespread adoption in consumer electronics, electric vehicles, and stationary energy storage systems. Continuous research and development efforts have focused on refining materials, optimizing cell design, and enhancing safety protocols to further improve performance and longevity.

Lead-Acid Batteries: A mature technology characterized by reliability and cost-effectiveness, lead-acid batteries have found their niche in automotive starting, lighting, and ignition (SLI) systems. Their robust construction and predictable performance make them suitable for applications demanding consistent power delivery. Additionally, they have gained traction in renewable energy systems as a dependable backup power source.

Nickel-Metal Hydride (NiMH) Batteries: Offering a higher energy density compared to their nickel-cadmium predecessors, NiMH batteries have gained prominence in hybrid electric vehicles and portable electronics. While they have been gradually replaced by lithium-ion batteries in many consumer applications, their ability to deliver high discharge rates and withstand temperature extremes has maintained their relevance in specific niches.

5.2.1.2 *Emerging Battery Technologies*

The quest for higher energy density, longer cycle life, and improved safety has propelled the development of cutting-edge battery technologies. Figure 5.1 shows the specific energy densities to gradually improve.

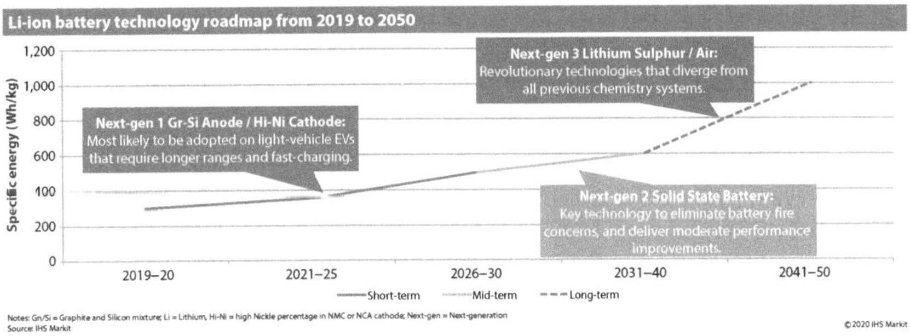

Figure 5.1 Specific energy densities to gradually improve as new battery technologies become ready for mass deployment.

Solid-State Batteries: By replacing the flammable liquid electrolyte with a solid material, solid-state batteries promise enhanced safety, energy density, and rapid charging capabilities. Overcoming challenges related to material selection, ion transport, and manufacturing processes is crucial for their commercialization.

Lithium-Sulfur Batteries: With a theoretical energy density significantly surpassing lithium-ion batteries, lithium-sulfur technology holds immense promise. However, addressing the challenges of sulfur dissolution, shuttle effects, and cycle life degradation remains essential for realizing its full potential.

Sodium-Ion Batteries: Leveraging the abundance of sodium, sodium-ion batteries offer a potentially lower-cost alternative to lithium-ion for large-scale energy storage applications. While their energy density currently lags behind lithium-ion, ongoing research aims to improve performance while maintaining cost advantages.

Flow Batteries: Unlike conventional batteries, flow batteries store energy in external liquid electrolytes, allowing for scalable capacity and long cycle life. Their suitability for grid-scale energy storage and industrial applications is evident, but their lower energy density compared to lithium-ion batteries presents a trade-off.

5.2.2 Supercapacitors

Supercapacitors, also known as ultracapacitors, are energy storage devices that excel in delivering high power in short bursts. Positioned between traditional capacitors and batteries, they combine the strengths of both while mitigating their respective weaknesses.

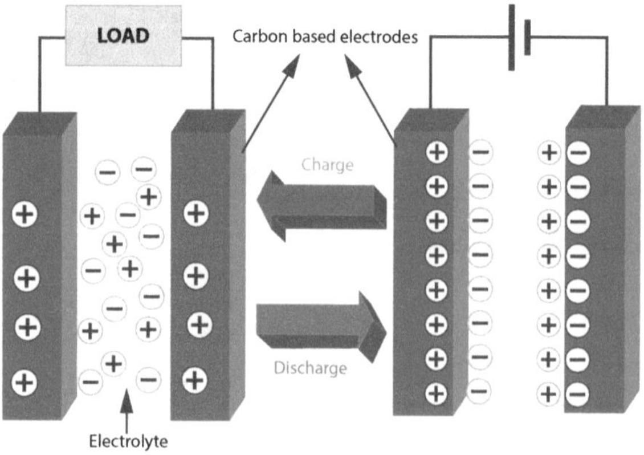

Figure 5.2 Supercapacitors.

Unlike batteries that rely on chemical reactions to store energy, supercapacitors store energy electrostatically. This is primarily achieved through two mechanisms: double-layer capacitance and pseudo capacitance. Figure 5.2 shows the double-layer capacitance arises from the accumulation of ions at the electrode-electrolyte interface, forming a double layer that stores electrical energy. Pseudo capacitance, on the other hand, leverages specific electrode materials like metal oxides and conducting polymers to store charge through rapid surface redox reactions.

This unique energy storage mechanism endows supercapacitors with several compelling attributes. Their exceptional power density allows them to deliver high currents rapidly, making them ideal for applications demanding short bursts of power. Moreover, supercapacitors boast an impressively long cycle life, capable of enduring countless charge-discharge cycles without significant performance degradation. This durability is a significant advantage over batteries, which tend to experience capacity fade over time.

Another key benefit is their rapid charging and discharging capabilities. Unlike batteries that may require several hours to fully charge, supercapacitors can be charged and discharged in a matter of minutes. This characteristic is particularly valuable in applications where quick energy replenishment is crucial. Furthermore, supercapacitors exhibit a wide operating temperature range, ensuring reliable performance in diverse environmental conditions.

Given their environmentally friendly nature, devoid of harmful materials, supercapacitors align with the growing emphasis on sustainable energy solutions. Their applications span a wide spectrum, including energy recovery systems, power backup, peak power delivery, portable electronics, and hybrid electric vehicles. In regenerative braking systems, for instance, supercapacitors efficiently capture kinetic energy during deceleration, storing it for subsequent use in acceleration.

Despite their numerous advantages, supercapacitors currently face limitations, primarily in terms of energy density compared to batteries. Overcoming this challenge while preserving their high power capabilities is a critical area of ongoing research. Scientists and engineers are actively exploring novel materials, electrode structures, and manufacturing processes to enhance energy storage capacity without compromising power delivery.

5.2.3 Fuel Cells

Fuel cells are electrochemical devices that directly convert chemical energy into electrical energy. Unlike batteries, which store a finite amount of energy, fuel cells produce electricity continuously as long as fuel and oxygen are supplied. This makes them a promising technology for various applications, from transportation to stationary power generation. Figure 5.3 shows the Fuel Cell Technology. At the heart of a fuel cell is an electrochemical process. Hydrogen is supplied to the anode, where it undergoes oxidation, releasing protons and electrons. The protons migrate through the electrolyte to the cathode, while the electrons travel through an external circuit, generating electricity. At the cathode, oxygen combines with the protons and electrons to form water [2].

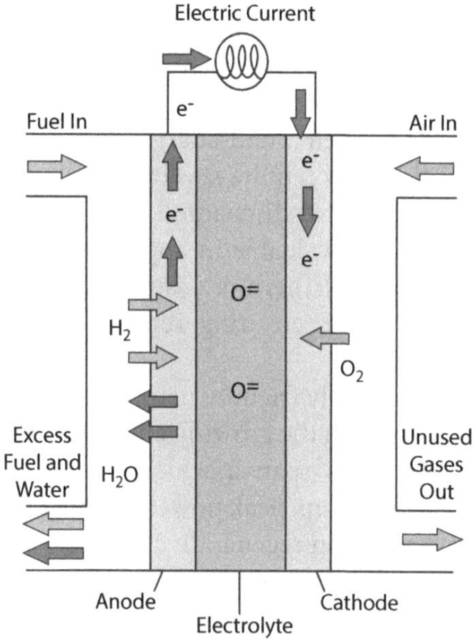

Figure 5.3 Fuel cell technology.

5.3 Materials for Batteries

5.3.1 Cathode Materials

The cathode, the positive electrode in a battery, is a critical component influencing overall performance, cost, and safety. Its composition significantly impacts energy density, power delivery, cycle life, and longevity.

Lithium Cobalt Oxide (LCO): Renowned for its high energy density and exceptional performance, LCO has been a staple in portable electronics. However, its relatively high cost, limited thermal stability, and cobalt supply chain concerns have spurred the exploration of alternatives [1].
Lithium Manganese Oxide (LMO): Offering a balance of cost-effectiveness and safety, LMO has gained traction in various applications. While its energy density is lower than LCO, its structural stability and abundance of manganese make it an attractive choice.
Lithium Iron Phosphate (LFP): Known for its exceptional safety and long cycle life, LFP has emerged as a preferred cathode material for electric vehicles and energy storage systems. Despite its lower energy density compared to LCO and LMO, its inherent stability and low cost have driven its widespread adoption.
Lithium Nickel Manganese Cobalt Oxide (NMC): This versatile cathode material allows for tuning of composition to optimize energy density, power density, and cycle life. By carefully balancing nickel, manganese, and cobalt content, manufacturers can tailor NMC cathodes for specific applications.
Lithium Nickel Cobalt Aluminum Oxide (NCA): Delivering high energy density, NCA cathodes are often found in high-performance electric vehicles. However, their susceptibility to thermal runaway necessitates advanced safety measures.

5.3.2 Anode Materials

The anode serves as the negative electrode within a battery and plays a pivotal role in determining overall cell performance, cost, and safety. During charge and discharge cycles, the anode undergoes reversible lithium-ion intercalation and deintercalation.

5.3.2.1 Graphite

Graphite has historically been the cornerstone of anode technology in lithium-ion batteries due to its low cost, abundant supply, and excellent electrical conductivity. Its layered structure facilitates efficient lithium-ion diffusion, resulting in commendable cycle life and safety profiles. However, its theoretical specific capacity limitations have spurred research into alternative materials capable of delivering higher energy densities [2].

5.3.2.2 *Advanced Anode Materials*

To enhance energy density and overall battery performance, researchers have explored a variety of anode materials:

Silicon: Possessing a theoretical specific capacity approximately tenfold that of graphite, silicon offers significant potential for boosting energy density. Nevertheless, substantial volumetric changes during lithiation lead to mechanical stress, particle fracture, and rapid capacity fade. To mitigate these challenges, researchers are actively investigating nanostructured silicon, alloying with other elements, and advanced electrode architectures.
Lithium Titanate Oxide (LTO): Renowned for its exceptional safety, rapid charging capabilities, and extended cycle life, LTO is employed in applications demanding high power and durability, such as electric buses and grid energy storage [11]. While its specific capacity is lower than graphite, its robust performance under demanding conditions makes it a valuable anode choice.
Graphene: Characterized by its two-dimensional structure, exceptional electrical conductivity, and expansive surface area, graphene holds promise as an anode material. It can potentially enhance lithium-ion diffusion kinetics, improve electrical conductivity, and bolster mechanical stability. However, challenges related to mass production and cost-effective synthesis must be addressed for widespread commercialization [3].
Alloy-Based Materials: Materials such as tin and aluminum alloys exhibit high theoretical capacities, but practical applications are hindered by severe volume changes during lithiation, leading to capacity loss and structural instability. To overcome these challenges, alloying with other elements, nanostructuring [5], and composite materials are being explored.

5.3.3 Electrolytes and Separators

Electrolytes and separators are crucial components that work in tandem to facilitate ion transport and prevent electrical short circuits within a battery.

5.3.3.1 *Electrolytes*

Electrolytes are ionically conductive media that enable the movement of lithium ions between the cathode and anode during charging and discharging [4]. They typically consist of a lithium salt dissolved in an organic solvent. Key properties of electrolytes include ionic conductivity, electrochemical stability, safety, and compatibility with other battery components.

5.3.3.2 Types of Electrolytes

Liquid Electrolytes: Commonly used in lithium-ion batteries, they offer high ionic conductivity but pose safety concerns due to flammability.
Gel Electrolytes: A combination of liquid and polymer electrolytes, they provide improved safety but often exhibit lower ionic conductivity.
Solid Electrolytes: These offer enhanced safety and potential for higher energy density but face challenges in terms of ionic conductivity and interfacial resistance.

5.3.4 Separators

Separators act as a physical barrier between the cathode and anode, preventing direct contact and electrical short circuits. They must allow for the passage of lithium ions while blocking the movement of electrons [13]. Key properties of separators include porosity, thickness, thermal stability, and mechanical strength.

5.3.4.1 Types of Separators

Microporous Polyolefin Separators: Commonly used in lithium-ion batteries, they offer a good balance of porosity, mechanical strength, and cost-effectiveness.
Ceramic Separators: Exhibiting excellent thermal stability and chemical resistance, ceramic separators are suitable for high-temperature applications.
Polymer Separators: These offer flexibility and good ionic conductivity but may have limitations in terms of mechanical strength.

5.4 Photovoltaic Materials

Photovoltaic (PV) materials are the heart of solar energy conversion, transforming sunlight directly into electricity. While crystalline silicon has been the industry standard for decades due to its reliable performance, its production energy intensity drives up costs. This has spurred the development of alternative materials and technologies [6].

Thin-film silicon, for instance, offers a compelling proposition with reduced material consumption and potential for flexible applications. However, its efficiency typically lags behind crystalline silicon. Cadmium telluride (CdTe) presents a promising alternative, boasting higher efficiency and lower production costs. Yet, the toxicity of cadmium remains a significant challenge. Copper indium gallium selenide (CIGS) is another contender, combining high efficiency with flexibility. However, the scarcity and cost of indium limit its widespread adoption.

In recent years, perovskite materials have emerged as a disruptive force in the PV industry. Their rapid efficiency gains and potential for low-cost manufacturing have generated immense excitement. However, stability and toxicity concerns must be addressed before large-scale commercialization can be realized.

To further enhance solar energy conversion, researchers are exploring tandem solar cells, which combine multiple materials to capture a broader spectrum of sunlight. This approach has the potential to significantly boost efficiency. Additionally, integrating PV cells with energy storage systems is essential to address the intermittent nature of solar power generation [12].

The environmental impact of PV material production and disposal is a critical concern. Minimizing the energy consumption of the manufacturing process, reducing waste, and developing effective recycling strategies are essential for a sustainable solar energy industry. Ensuring a stable supply of critical raw materials through responsible sourcing and recycling is also crucial.

5.5 Hydrogen Storage Materials

5.5.1 Metal Hydrides

Metal hydrides, substances formed when hydrogen atoms are absorbed into the crystalline structure of a metal alloy, offer a compelling avenue for energy storage due to their reversible hydrogenation and dehydrogenation properties. This capability renders them suitable for a spectrum of energy applications, including hydrogen storage for fuel cell vehicles, stationary energy storage, thermal energy storage, and battery technology [7]. A cornerstone of metal hydride-based energy storage systems lies in the selection of appropriate materials. Magnesium (Mg) and its alloys, renowned for their high hydrogen storage capacity and cost-effectiveness, have been extensively investigated. However, their practical application is hindered by sluggish hydrogen sorption kinetics and the demand for elevated temperatures to release hydrogen. To address these challenges, researchers have delved into alloying Mg with elements such as nickel, iron, and cobalt to enhance hydrogen absorption/desorption rates and reduce operating temperatures.

Lanthanum nickel (LaNi5) and its derivatives have garnered attention for their rapid hydrogen exchange kinetics and commendable cycle life. Their integration into nickel-metal hydride (NiMH) batteries underscores their viability in energy storage systems. Nevertheless, their hydrogen storage capacity is comparatively lower than that of magnesium-based hydrides. Titanium (Ti) and its alloys, celebrated for their stability and corrosion resistance, find utility in hydrogen storage tanks and as integral components in certain metal hydride batteries. Complex hydrides, characterized by their intricate structures comprising metal cations and complex hydride anions, such as sodium alanate (NaAlH4) and lithium borohydride (LiBH4), exhibit exceptionally high theoretical hydrogen storage capacities. Yet, their practical implementation is impeded by stability concerns and sluggish kinetics. Overcoming the inherent challenges of metal hydrides, such as sluggish kinetics, limited hydrogen storage capacity, and durability issues, is paramount for their widespread adoption. Ongoing research endeavors are focused on refining materials, optimizing synthesis processes, and exploring innovative system designs. Advancements in nanotechnology, catalysis, and materials science hold the promise of unlocking the full potential of metal hydrides [19].

In the realm of energy storage, metal hydrides present a multifaceted opportunity. Their capacity to store hydrogen efficiently enables their integration into fuel cell vehicles, extending driving ranges and reducing dependence on fossil fuels. Moreover, metal hydride-based systems can serve as stationary energy storage solutions, capturing renewable energy surplus for subsequent utilization during periods of peak demand. The thermal energy released or absorbed during the hydrogenation/dehydrogenation process can be harnessed for heating and cooling applications, contributing to energy efficiency. Additionally, nickel-metal hydride batteries, which leverage metal hydrides as the negative electrode, offer a mature and reliable energy storage technology [14]. As research progresses and technological advancements materialize, metal hydrides are poised to play an increasingly pivotal role in shaping the energy landscape [10]. Their potential to store and release hydrogen efficiently, coupled with their versatility in various applications, positions them as a promising solution to address the growing energy storage challenges of the 21st century.

5.5.2 Chemical Hydrides

Chemical hydrides, particularly complex hydrides, represent a compelling avenue for high-density energy storage. These compounds, composed of metal cations and complex hydride anions, exhibit remarkable potential in storing hydrogen, a clean and versatile energy carrier. A prime example is sodium alanate (NaAlH4), which boasts an exceptionally high theoretical hydrogen storage capacity. However, the practical realization of this potential is hindered by significant challenges [8]. The release and absorption of hydrogen typically necessitates elevated temperatures and pressures, limiting its applicability. Moreover, the inherent instability of many complex hydrides under ambient conditions poses additional obstacles.

To unlock the full potential of chemical hydrides, intensive research is directed towards overcoming these hurdles. Catalyst development is paramount in accelerating hydrogen exchange kinetics at lower temperatures and pressures. Simultaneously, the pursuit of novel hydride materials with enhanced stability and kinetics is a focal point. Nanostructuring techniques are also explored to augment surface area and reaction rates [21].

Beyond hydrogen storage, chemical hydrides offer versatility in energy applications. Their exothermic and endothermic properties during hydrogenation and dehydrogenation, respectively, enable their integration into thermal energy storage systems. Furthermore, their potential as anode materials in advanced battery technologies is being investigated [20].

While substantial progress has been made, the commercialization of chemical hydrides for energy storage remains a complex endeavor [15]. The intricate interplay of materials science, chemistry, and engineering is essential to develop practical and cost-effective solutions. As research advances and technological barriers are surmounted, chemical hydrides have the potential to become a cornerstone of a sustainable energy future.

However, it is crucial to acknowledge that the path forward is fraught with challenges. Issues such as hydrogen purity, material degradation, and system integration must be meticulously addressed. Nevertheless, the potential rewards in terms of energy density, safety, and environmental impact justify continued research and development efforts in this promising field.

5.5.3 Carbon-Based Materials

Carbon-based materials have emerged as indispensable components in the burgeoning field of energy storage, owing to their unique amalgamation of properties. Their inherent attributes, such as exceptional electrical conductivity, substantial surface area, and robust chemical stability, render them ideal candidates for a spectrum of energy storage applications [19].

Graphene, a two-dimensional allotrope of carbon, stands as a prime exemplar of these materials. Its extraordinary electrical conductivity, coupled with its vast surface area, empowers it to significantly enhance the performance of both batteries and supercapacitors. As an anode material in lithium-ion batteries, graphene contributes to improved energy density, power delivery, and cycle life. Moreover, its incorporation into supercapacitors leads to augmented charge storage capacity, resulting in higher power densities and expedited charging/discharging rates [2].

Carbon nanotubes (CNTs), characterized by their cylindrical structure and exceptional electrical conductivity, are another pivotal carbon-based material in energy storage. Their integration into battery electrodes can bolster rate capability and cycle life, while their utilization in supercapacitors can enhance energy density.

Activated carbon, renowned for its porous structure and substantial surface area, has been extensively employed as an electrode material in supercapacitors. This material's ability to adsorb ions onto its surface facilitates rapid charge storage and discharge, making it a cornerstone in high-power applications.

Carbon aerogels, distinguished by their ultralight and highly porous nature, offer promising prospects for energy storage. Their exceptional electrical conductivity and thermal stability make them attractive candidates for both supercapacitors [9] and battery electrodes.

Beyond their application as electrode materials, carbon-based materials also play a crucial role in other energy storage components. For instance, they serve as conductive additives in battery electrodes, facilitating electron transport and enhancing overall performance. In fuel cells, carbon-based materials function as catalyst supports, augmenting the efficiency and durability of the catalyst.

Despite the numerous advantages offered by carbon-based materials, several challenges persist. Mass production at a commercially viable scale, cost-effective synthesis, and performance optimization remain key areas of research and development. To address these challenges, scientists and engineers are actively exploring novel approaches, such as doping, functionalization, and hybridization with other materials, to tailor the properties of carbon-based materials for specific energy storage applications [16].

The integration of carbon-based materials with other advanced materials, including metal oxides and conducting polymers, presents an exciting avenue for creating hybrid energy storage devices that combine the strengths of different materials. Such synergistic combinations hold the potential to deliver unprecedented performance metrics.

Moreover, the development of sustainable and environmentally friendly methods for the production and recycling of carbon-based materials is imperative for the long-term viability of this technology [8]. By addressing these challenges and capitalizing on the inherent advantages of carbon-based materials, researchers can pave the way for a new era of high-performance and sustainable energy storage solutions.

5.6 Challenges and Future Directions

The realization of a sustainable energy future hinges on the development of advanced materials for energy storage and conversion. However, significant hurdles persist. Achieving optimal energy and power density while maintaining safety and cycle life remains a formidable challenge. Cost-effective and scalable manufacturing processes are essential for widespread adoption and commercial viability. Moreover, the environmental impact of material extraction, production, and disposal necessitates a concerted focus on sustainability.

To navigate these complexities, a multidisciplinary approach is imperative. The discovery of novel materials with tailored properties, coupled with advanced characterization techniques, is crucial for understanding material behavior and performance. Computational modeling and simulation can accelerate the materials discovery process by predicting properties and optimizing compositions. Additionally, nanotechnology offers opportunities to engineer materials with enhanced functionalities [17].

Recycling and reuse strategies are vital for minimizing waste and reducing reliance on virgin materials. The development of circular economy models for energy storage materials is essential to ensure long-term sustainability. Furthermore, addressing supply chain vulnerabilities and securing a stable supply of critical materials is crucial for the industry's resilience.

Specific focus areas include the pursuit of solid-state electrolytes, post-lithium-ion chemistries, and high-nickel cathodes for batteries. In fuel cells, improving catalyst efficiency, durability, and cost-effectiveness remains paramount. For supercapacitors, enhancing energy density while preserving high power density is a key challenge. Thermal energy storage necessitates the development of high-performance phase change materials and efficient thermal management systems [18].

5.7 Conclusion

The pursuit of clean, efficient, and sustainable energy systems is inextricably linked to advancements in materials science. The development of innovative materials for energy storage and conversion is pivotal to achieving this goal. From the high-energy density demands of electric vehicles to the grid-scale challenges of renewable energy integration, materials play a critical role. While significant progress has been made in areas such as lithium-ion batteries, fuel cells, and solar cells, numerous challenges persist. Enhancing energy density, power density, cycle life, and safety while reducing costs and environmental impact remain key objectives. The discovery and optimization of new materials, coupled with innovative device designs, are essential to overcome these hurdles. A multidisciplinary approach, encompassing materials science, chemistry, physics, and engineering, is crucial for driving advancements in this field. Collaboration between academia, industry, and government is vital to accelerate research and development. Additionally, a strong emphasis on sustainability, including material sourcing, production, and recycling, is imperative to ensure the long-term viability of energy storage and conversion technologies. The future of energy storage and conversion is bright, with promising developments on the horizon. Solid-state batteries, perovskite solar cells, and advanced hydrogen storage materials offer exciting possibilities. However, the path to a sustainable energy future requires continued investment in research, development, and commercialization of these technologies. By harnessing the power of materials science, we can pave the way for a cleaner, more resilient, and energy-independent world.

References

1. Hu, X., *et al.*, A Review of Second-Life Lithium-Ion Batteries for Stationary Energy Storage Applications. *Proc. IEEE*, 110, 6, 735–753, June 2022, doi: 10.1109/JPROC.2022.3175614.
2. Vesely, L., Syblik, J., Entler, S., Stepanek, J., Zacha, P., Dostal, V., Optimization of Supercritical CO2 Power Conversion System With an Integrated Energy Storage for the Pulsed DEMO. *IEEE Trans. Plasma Sci.*, 48, 6, 1715–1720, June 2020, doi: 10.1109/TPS.2020.2971718.
3. Mugyema, M., Kamper, M.J., Wang, R.-J., Sebitosi, B.A., Performance and Cost Comparison of Drive Technologies for a Linear Electric Machine Gravity Energy Storage System. *IEEE Access*, 12, 46953–46966, 2024, doi: 10.1109/ACCESS.2024.3383161.
4. Jin, J., *et al.*, Improving High-Temperature Energy Storage Performance of Silicon-Integrated Oxide Film Capacitors via Inserting a Graphene Buffer Layer. *IEEE Electron Device Lett.*, 42, 8, 1216–1219, Aug. 2021, doi: 10.1109/LED.2021.3087024.

5. Li, G., *et al.*, The Interaction Between a High-Temperature Superconducting Coil and In-Series Permanent Magnets. *IEEE Trans. Appl. Supercond.*, 34, 3, 1–5, May 2024. Art no. 5700505, doi: 10.1109/TASC.2024.3356484.
6. Kamaei, S., Ghini, M., Gilani, A., Gastaldi, C., Collette, E., Ionescu, A.M., Ferroelectric Supercapacitors by Combining Polarization Switching and Negative Capacitance Effects for On-Chip Energy Storage. *IEEE Electron Device Lett.*, 44, 9, 1579–1582, Sept. 2023, doi: 10.1109/LED.2023.3296945.
7. Nel, B.A.P., Skrivervik, A.K., Gustafsson, M., Stored Energies and Q-Factor Expressed in Material Derivatives. *IEEE Antennas Wirel. Propag. Lett.*, 23, 1, 19–23, Jan. 2024, doi: 10.1109/LAWP.2023.3314859.
8. Wu, M., *et al.*, Temperature Stability of Dielectric Constant and Energy Storage Properties of (Pb1-x,Lax)(Zr0.65,Ti0.35)O3 Relaxor Ferroelectric Thin Films. *IEEE Trans. Dielectr. Electr. Insul.*, 28, 6, 2052–2057, December 2021, doi: 10.1109/TDEI.2021.009818.
9. Fang, S., Lv, Z., Chao, G., Methods of Increasing the Energy Storage Density of Superconducting Flywheel. *IEEE Trans. Appl. Supercond.*, 31, 8, 1–5, Nov. 2021, doi: 10.1109/TASC.2021.3094450. Art no. 5700705.
10. Zan, Y., Fang, J., Pei, X., Liao, X., Wang, X., Zeng, X., Influence of Structure Parameters of Flux Diverters on Performance of Superconducting Energy Storage Coil. *IEEE Trans. Appl. Supercond.*, 33, 8, 1–8, Nov. 2023, doi: 10.1109/TASC.2023.3304606. Art no. 5700508.
11. Farivar, G.G., *et al.*, Grid-Connected Energy Storage Systems: State-of-the-Art and Emerging Technologies. *Proc. IEEE*, 111, 4, 397–420, April 2023, doi: 10.1109/JPROC.2022.3183289.
12. Mohiti, M., Mazidi, M., Oggioni, N., Steen, D., Tuan, L.A., An IGDT-Based Energy Management System for Local Energy Communities Considering Phase-Change Thermal Energy Storage. *IEEE Trans. Ind. Appl.*, 60, 3, 4470–4481, May-June 2024, doi: 10.1109/TIA.2024.3359126.
13. Wang, T., Ding, K., Du, S., Zou, C., Study on Electromagnetic Stress Optimization and its Resistant Structure in Toroidal HTS Magnet Applied in High-Energy Storage Density SMES. *IEEE Trans. Appl. Supercond.*, 31, 8, 1–5, Nov. 2021, doi: 10.1109/TASC.2021.3107822. Art no. 4902305.
14. Wang, X., Jiang, Z., Chen, X., Han, X., Hao, X., Tang, X., Enhanced Energy-Storage Performance of an All-Inorganic, Antiferroelectric, Thin-Film via Orientation Adjustments. *IEEE Access*, 8, 217246–217254, 2020, doi: 10.1109/ACCESS.2020.3039349.
15. Costello, B. and Davis, J.A., Breakdown Field Strength Variations and Energy Density Limits of Nanoparticle Composite Materials. *IEEE Trans. Nanotechnol.*, 19, 811–819, 2020, doi: 10.1109/TNANO.2020.3037897.
16. Rouholamini, M., *et al.*, A Review of Modeling, Management, and Applications of Grid-Connected Li-Ion Battery Storage Systems. *IEEE Trans. Smart Grid*, 13, 6, 4505–4524, Nov. 2022, doi: 10.1109/TSG.2022.3188598.

17. Wu, H., Shan, C., Fu, S., *et al.*, Efficient energy conversion mechanism and energy storage strategy for triboelectric nanogenerators. *Nat. Commun.*, 15, 6558, 2024, https://doi.org/10.1038/s41467-024-50978-7.
18. Hellerschmied, C., Schritter, J., Waldmann, N., *et al.*, Hydrogen storage and geo-methanation in a depleted underground hydrocarbon reservoir. *Nat. Energy*, 9, 333–344, 2024, https://doi.org/10.1038/s41560-024-01458-1.
19. Gurnani, R., Shukla, S., Kamal, D., *et al.*, AI-assisted discovery of high-temperature dielectrics for energy storage. *Nat. Commun.*, 15, 6107, 2024, https://doi.org/10.1038/s41467-024-50413-x.
20. Chen, S., Ye, Z., Meng, Y., Optimization clearing strategy for multi-region electricity-heat market considering shared energy storage and integrated demand response. *Sci. Rep.*, 14, 21368, 2024, https://doi.org/10.1038/s41598-024-72397-w.
21. Yang, F., Peng, P., Yan, Z.Y., *et al.*, Vertical iontronic energy storage based on osmotic effects and electrode redox reactions. *Nat. Energy*, 9, 263–271, 2024, https://doi.org/10.1038/s41560-023-01431-4.

6

Biogenic AgNPs: Leaf-Mediated Green Synthesis, Analytical Spectroscopic Characterization, and Applications

Susanta Das

School of Engineering, Ajeenkya DY Patil University, Pune, Maharashtra, India

Abstract

Metal nanoparticles (NPs), synthesized from noble metals, have been investigated for countless applications because of their large surface areas (shapes and sizes), and their unique chemical, biological, and physical properties. Silver nanoparticles (AgNPs) have demonstrated exceptional and remarkable antimicrobial and catalytic activities. AgNPs can be synthesized using conventional/traditional (physical and chemical) and green approaches. However, green synthesis techniques have some advantages over conventional methods. In particular, leaf-assisted biogenic synthesis of AgNPs, within the realm of green synthesis, has evolved into one of the most promising techniques for the fabrication of non-toxic, stable, cost-effective, biocompatible, and eco-friendly AgNPs. In this chapter, we have reviewed leaf-assisted green synthesis of AgNPs, their spectroscopic characterizations employing various analytical techniques, and their promising applications in diverse fields.

Keywords: Metal NPs, various medicinal plants, leaf extract, biosynthesis, biogenic AgNPs, characterizations, applications

6.1 Introduction

Metal NPs such as silver, gold, copper, and platinum NPs synthesized from noble metals, have undergone extensive investigation both from

Email: susanta2024das@gmail.com

Kolla Bhanu Prakash, S. V. Ranganayakulu and K. S. Jagannatha Rao (eds.) New Frontiers in Materials Science: Interdisciplinary Approaches to Innovation and Technologies, (97–118) © 2025 Scrivener Publishing LLC

fundamental and technological application viewpoints due to their large surface areas (shapes and sizes) and distinctive properties [1–11].

The broad spectrum of beneficial characteristics of AgNPs over other NPs has propelled them to the forefront of diverse fields. Consequently, the scientific community has shown escalating interest in AgNPs due to their multifaceted applications across various sectors [12–22]. In particular, AgNPs with varying sizes (1–100 nm) have gained substantial attention. Their applications include anticancer, antimicrobial (antibacterial and antifungal), antioxidant, agriculture, food production and preservation, packaging, cosmetics, water purification, painting/ coating, textiles, biosensing technology, nanodrug carrier, catalytic capabilities, ability to promote wound and pain healing activities, and many more [12–22].

AgNPs can be synthesized through traditional (physical and chemical) and green approaches [2, 3, 12, 15, 19, 20]. However, traditional/ conventional techniques require the use of hazardous chemicals for synthesis [21]. Moreover, these methods are costly and involve complex steps and parameters (pressure, temperature, etc.) to control the growth and interaction kinetics and achieve the desired stability, morphology (structure, shape, and size), and properties (biocompatibility, etc.) of the AgNPs, which are crucial for their intended applications [11, 14, 15, 17, 19, 22]. In addition to economic and technical challenges, physically and chemically synthesized AgNPs pose significant risks to the entire ecosystem [11, 21, 23]. If the release of AgNPs from various sources, such as during production, contaminated water, or household units like antibacterial coatings/painting, exceeds the tolerance level, they can have significant adverse effects on humans, animals, plants, and the environment [23]. To address these concerns, scientists have turned to green synthesis (biosynthesis or green chemistry), a biological approach that offers less complex, environmentally friendly, and cost-effective alternative while ensuring safety [14]. Furthermore, green-synthesized AgNPs demonstrate increasing stability and biocompatibility with cells and tissues as particle size decreases compared to those synthesized through physical and chemical routes [19–22]. Consequently, the green synthesis technique has emerged as a promising method for developing AgNPs in recent years [8, 11, 13, 23].

Medicinal plants are sources of various natural chemicals and metabolites/biocompounds (e.g., phenolic and flavonoids) [24–26]. These compounds are useful in treating numerous health-related disorders/problems and diseases. A few of them are oxidant, allergy, diabetes, inflammatory, ulcer, arthritic, osteoporosis, wound, pain, burn, fever, and cancer [24–26]. Thus, the fabrication of AgNPs using plants is a useful technique among the various green synthesis methods using biological systems (e.g., bacteria, yeast, fungi, algae, plants, etc.) and compounds (such as carbohydrates) [26, 27].

In this chapter, we have focused on the biofunctionalized AgNPs from the extracts of leaves of various medicinal plants, their characterization, and potential applications in healthcare as well as controlling environmental pollution.

6.2 Synthesis, Characterization, and Applications of Biogenic AgNPs

We have reviewed several articles on the leaf-mediated synthesis of AgNPs, their characterization, and potential applications in nanomedicine and environmental sustainability. In this section, we have briefly discussed them.

Anandalakshmi *et al.* [28] employed a green synthesis method to prepare biocompatible AgNPs with various medicinal properties using the leaf extract of the *Pedalium murex* plant. The experimental procedure involved collecting, cleaning, and drying fresh leaves, followed by their pulverization into a powder. A mixture of appropriate proportions of powder and distilled water was boiled for a short duration, and filtered to obtain the leaf extract. It was further stored at low temperatures for eco-friendly synthesis purposes. The leaf extract and low-molarity $AgNO_3$ solution (0.01 mM) were mixed. The reduction of Ag^+ ions into neutral Ag and the formation and capping of stable AgNPs were confirmed by the color change of the mixture (light yellow to dark brown) due to surface plasmon resonance (SPR). Different concentrations of leaf extracts were prepared to evaluate the antibacterial activities of synthesized NPs against various bacterial species. UV-Vis spectra of biosynthesized AgNPs displayed shifts in SPR peak (430 to 424 nm), implying the involvement of different compounds present in leaf extract in the reduction of silver ions and formation of biogenic NPs. Photoluminescence (PL) spectra also exhibited peak shifts, from around 478 nm to 650 nm. Fourier transform infrared (FTIR) measurements confirmed the presence of various biochemicals (flavonoids, alkaloids, steroids, and proteins) in extract through distinct peaks at different wavelengths. These peaks were attributed to the vibration and stretching of bonds (C=C, C-N, O-H, etc.) associated with different molecules in the leaf extract. X-ray diffraction (XRD) pattern unveiled the formation of *fcc* crystals with various plane orientations, characterized by an average size of ~14 nm and a lattice constant of ~4 nm. Peaks corresponding to different chemicals were also detected in the X-ray spectra. Field emission scanning electron microscope (FESEM) images showcased spherically shaped NPs with sizes ranging from 20 nm to 50 nm. Energy dispersive x-ray analysis (EDAX) corroborated the presence of various elements (e.g., Ag, O, C, Cl, etc.) in the synthesized AgNPs. Transmission electron microscope (TEM) images further revealed spherical NPs with an average size of ~50 nm. DLS experiments indicated a wide size distribution (10–150 nm), with an average size of ~73 nm. Zeta potential demonstrated a peak at ~-8 mV, suggesting repulsion among the negatively charged and dispersed AgNPs due to the diverse biocompounds present in the leaf extract of the *Pedalium murex* plant. The antibacterial properties of NPs at different concentrations were measured against various gram-positive (*Bacillus subtilis, Staphylococcus aureus, Micrococcus flavus*, and *Bacillus pumilus*) and gram-negative (*Escherichia coli, Klebsiella pneumoniae*, and *Pseudomonas aeruginosa*) bacteria using the disc diffusion method. The study observed significant inhibition zones based on the concentrations and bacterial types, suggesting promising applications of these NPs in wound and cancer treatment, as well as antibacterial activities. The researchers also endorsed the utilization of these AgNPs across diverse industrial sectors.

Brassica oleracea (cauliflower), *Brassica oleracea capitata* (cabbage), *Raphanus sativus* (radish) vegetables, all belong to the same family of Brassicaeae, are considered rich sources of various types of minerals, vitamins, fibers, bioactive compounds (e.g., flavonoids), antioxidants, and various other nutritious elements, and are therefore, considered for their antibacterial efficacies, developing immunity, and detoxication [29–31].

Ansar and his group [29] reported the synthesis of stable and biocompatible AgNPs by mixing leaf extract of *B. oleracea* with an $AgNO_3$ solution, highlighting their potential applications in various fields. The process involved thoroughly cleaning and drying fresh *B. oleracea* leaves, boiling approximately 20 g of these leaves for a period of time, and then allowing the extract to cool. A small portion of the extract and 1 mM of $AgNO_3$ were mixed under normal conditions. After a certain duration, the mixture underwent a color change (greenish-yellow to brown), confirming the reduction of Ag^+ ions to Ag^0, as well as the capping and formation of stable AgNPs through various compounds of leaf extract. UV-Vis spectra of AgNPs at different concentrations exhibited a sharp absorption peak at ~415 nm, attributed to SPR, which further confirmed the biosynthesis of AgNPs assisted by leaf extract. TEM images revealed the production of uniformly distributed spherical AgNPs with an average diameter of approximately 20 nm. EDX measurements disclosed the presence of various elements, with Ag (at 3 keV) being the most prominent, indicating that other elements such as oxygen and chlorine played roles as capping agents in this green synthesis method. FTIR results exhibited absorption peaks at ~3500 and 1600 cm^{-1}, demonstrating the presence of O-H and C-H functional groups in the NPs. Antibacterial properties of NPs were assessed against both gram-positive and gram-negative bacteria (*Bacteroides fragilis, Staphylococcus epidermidis, Enterococcus faecalis, Proteus mirabilis, Pseudomonas aeruginosa, Escherichia coli, Klebsiella pneumoniae*, and *Bacteroides fragilis*) with varying inhibition zones (ranging from 9 to 14 mm). The highest inhibition zones were observed for gram-positive *Staphylococcus epidermidis* (~14 mm) and gram-negative *Pseudomonas aeruginosa* (~12 mm). Furthermore, the cytotoxicity of AgNPs against cancer cells (MCF-7) was found to be highly effective, with effectiveness increasing as the concentration of NPs increased. The researchers also evaluated the antioxidant properties of AgNPs against 2,2-diphenyl-1-picrylhydrazyl (DPPH), nitric oxide, superoxide, and hydroxyl radicals. They observed that the scavenging activities of biocompatible AgNPs increased with higher concentrations of AgNPs. Results recommended that these NPs can act as promising anticancer, antibacterial, and antioxidant agents.

Tamileswari *et al.* [30] employed *Brassica oleracea* (cauliflower) and *Brassica oleracea capitata* (cabbage) as reduction and capping agents to synthesize stable AgNPs from $AgNO_3$. Both cauliflower and cabbage, 25 g each, were cleaned, dried, and cut into pieces. They were separately boiled for a short duration. The resulting solutions were filtered to obtain the cauliflower and cabbage extracts. A specific amount of each extract was mixed with $AgNO_3$ solution, and the mixtures were kept at ambient conditions. Within 15 minutes, the mixture colors changed to dark brown (cauliflower) and light brown (cabbage), indicating the role of biocompounds present in vegetables in reducing Ag^+ ions and forming stable AgNPs. The samples were further processed for purity. UV-Vis spectra showed SPR peaks at ~420 and 428 nm (cauliflower and cabbage), resulting from electron–photon interactions. XRD data indicated *cubic* and *fcc* structures for the two samples with various sizes (36–42 nm). SEM images displayed flake and spherical shapes of AgNPs of different sizes (30–50 nm). EDAX elemental testing confirmed the presence of Ag through a peak of ~3 keV. Antibacterial properties were evaluated on four bacteria types, both gram-positive and gram-negative (*Klebsiella pneumoniae, Bacillus subtilis, Staphylococcus aureus*, and *Escherichia coli*). Effective inhibition zones were observed, with the largest (16 mm) for *Bacillus subtilis*. This study highlighted the potential of cauliflower and cabbage extracts as eco-friendly and cost-effective agents for environmentally benign AgNPs with antibacterial applications.

Singh *et al.* [31] used nine North Indian variants of the *Brassicaeae* family to synthesize highly stable AgNPs under different chemical conditions and parameters. They characterized these NPs using different techniques and compared their antibacterial properties with AgNPs synthesized using a chemical route and those synthesized from other (South Indian) species of the same *Brassicaeae* family. The nine North Indian variants consisted of six from *Brassica oleracea* (cauliflower) (Ashwini, Deepali, Kartik Sankar, Meghna, Paushja, and Sharad) and three from *Raphanus sativus* (radish) (Himani, Gulabi, and Jamuni), which were used for experimentation. Extracts from the tissues were prepared and maintained at a low temperature for synthesis. $AgNO_3$ solutions of different concentrations, volumes, and pH levels were prepared under different temperatures to assess the reduction effect of extracts on the solution and fabrication of AgNPs. Additionally, AgNPs were chemically synthesized by combining $AgNO_3$ solution and trisodium citrate ($Na_3C_6H_5O_7$) under controlled chemical conditions. Confirmation of AgNPs fabrication was achieved when the colorless mixtures exhibited a yellowish-brown hue in the green synthesis method and a gold color in the chemical synthesis method, respectively. UV-Vis spectra of the green-synthesized AgNPs showed characteristic SPR peaks between 400 nm and 435 nm for both cauliflower and radish, indicating the bioreduction capabilities of secondary metabolites in the plant extract to reduce the Ag^+ ions into stable AgNPs. The effects of controlled parameters [pH, time, and temperature (room temperature, 40°C, 60°C, and 80^{v}C)] on the eco-friendly AgNPs were further measured to understand the bioreduction times, outputs, and dispersities of AgNPs. TEM and nanoparticle tracking analysis (NTA) measurements revealed the formation of spherical NPs of various sizes (4 to 18 nm), depending on species types through the green synthesis method. The values were smaller in comparison to both chemically synthesized AgNPs and green synthesized AgNPs from the South Indian variants of the same *Brassicaeae* family. XRD data supported the formation of biofunctionalized AgNPs of *fcc* structures for both cauliflower and radish extracts. Fourier transform infrared spectroscopy (FTIR) spectra of both cauliflower and radish synthesized AgNPs displayed various absorption peaks, indicating the substantial presence and stretching of various bonds (C=C, C=O, S=O, O-H, C-H, etc.) associated with different bioactive molecules (such as benzenes, alkenes, amines, proteins, hydroxyls, carbonyls, and phenolics) of extracts, further signifying the roles of these metabolites (functional groups) in reduction/capping/formation of highly stable NPs. Moreover, UV-Vis spectra of cauliflower and radish extracts confirmed the presence of these bioreductants, suggesting their sustainable presence in the AgNPs and their active participation in the synthesis process. Antioxidant properties of AgNPs on DPPH varied between cauliflower and radish extracts, suggesting the presence of varying amounts of different free radical scavengers (phytochemicals) (e.g., phenolics and flavonoids) in the extracts. The antibacterial efficiencies of NPs of diverse concentrations were investigated on different bacteria, including *Kocuria* and *Promicromonospora* (gram-positive) and *Escherichia coli, Myroides,* and *Pseudomonas aeruginosa* (gram-negative). As the antibacterial efficacy of NPs depends on their size, shape, and structure (smaller size enhances antibacterial efficacy), the AgNPs synthesized from North Indian variants of cauliflower and radish extracts exhibited better performance due to their smaller sizes compared to those chemically synthesized and South Indian variants of cauliflower and radish extracts. Moreover, the antibacterial efficiencies of NPs from the extracts of cauliflower and radish varied among different variants and different bacteria. The results further revealed that the antibacterial properties of NPs did not vary significantly with concentrations of $AgNO_3$/NPs.

Catharanthus roseus (C. roseus) is renowned for its various medicinal properties, including antibacterial, antifungal, antibiotic, antioxidant, and antiviral effects, owing to the existence of numerous beneficial biochemicals (flavonoids and alkaloids) [32].

Ponarulselvam *et al.* [33] prepared AgNPs from the leaf extract of *C. roseus* utilizing a simple green-chemistry method without employing any hazardous chemical agents. They mixed pieces of 10 g of fresh leaves of *C. roseus* with 100 mL of refined water and boiled it for a few minutes. The solution was then filtered to produce leaf extract. Subsequently, the extract and $AgNO_3$ solution were mixed at controlled conditions. The mixture exhibited a transition of color (colorless to brown-yellow). This indicated the formation of stable biogenic AgNPs through secondary metabolites of the leaves of *C. roseus* plant. EDX results confirmed this phenomenon. UV-Vis measurements of NPs as a function of temperature revealed two peaks at 400 nm and 410 nm with the highest intensities. TEM images showed the formation of spherical NPs of various diameters (35–55 nm). XRD patterns confirmed the formation of an *fcc* crystal structure. The researchers further investigated the antimalarial effects on the malaria parasite (*P. falciparum*) using different concentrations of *C. roseus* leaf-extracted AgNPs, observing the formation of noticeable inhibition zones around these NPs based on concentration. These results underscored the promising nanobiotechnological utilizations of these NPs in many industries.

Alkammash [34] reported AgNPs utilizing leaf extracts from *Artemisia sieberi* (family: *Asteraceae*) and *Calotropis procera* (family: *Asclepiadaceae*) plants, which are known for their various traditional medicinal applications.

The extracts were prepared by employing standard cleaning, drying, and boiling methods on fresh leaves of *Artemisia sieberi* and *Calotropis procera*. The extracts were then mixed with 100 mL of $AgNO_3$ solutions at room temperature and subsequently centrifuged to synthesize two distinct types of stable NPs. The formation of NPs was established by a color change (yellow to brown), which indicated the effective roles of biomolecules (phytochemicals) from extracts in the reduction and formation of NPs. The UV-Vis absorption spectrum further validated it through a peak (responsible for the color change) at around 480 nm. SEM images uncovered various shapes, predominantly spherical, with varying sizes (8–20 nm). The researcher further asserted that all the results are similar to published values.

Syzygium jambos (*S. jambos*) (family: Myrtaceae) is renowned for its abundant phenolic and flavonoid compounds, such as gallic acid, rutin hydrate, and chlorogenic acid [35]. These compounds are found in various parts (leaves, fruits, seeds, and bark) of the plant. As a result, *S. jambos* has long been recognized for its diverse traditional medicinal applications, encompassing antibacterial, antifungal, cytotoxicity, anti-inflammatory properties, and nutritional benefits.

Dutta *et al.* [36] investigated the viability of Ag and Au NPs derived from S. *jambos* leaf and bark extracts as medicinal agents against malaria (*Plasmodium falciparum*). They employed rapid, easy, and environmentally friendly biosynthesis and chemical routes to synthesize spherical NPs. The researchers characterized these AgNPs using techniques like UV-Vis spectroscopy, zeta potential, TEM, XRD, nuclear magnetic resonance (NMR), and FTIR. Interestingly, the average size of NPs extracted from the leaves was approximately 5 nm, while for NPs from the bark, it ranged from 8 to 10 nm. The study revealed that this biosynthesis approach yielded AgNPs and AuNPs with high antimalarial properties, making them potential candidates for treating *Plasmodium falciparum*, the causative agent of malaria. These findings highlight the promising applications of *S. jambos*-derived NPs as effective antimalarial agents.

Thampi *et al.* [37] utilized the leaf extract of *Syzygium samarangense* (*S. samarangens*) (Wax Apple) (family: *Myrtaceae*) as the reducing agent for the green fabrication of AgNPs from $AgNO_3$ reagent. Fresh leaves from *S. samarangense* were processed to form a powder, which was then mixed with distilled water to create a solution. The solution was heated to 80°C for a period and subsequently cooled before being filtered. The resulting leaf extract was kept at a low temperature to preserve its bioreduction quality for the synthesis of NPs. The extract and $AgNO_3$ solution were mixed at appropriate proportions at a temperature of around 80°C for an hour.

The mixture changed color (light yellow → dark brown), announcing the formation of AgNPs. They were then purified and dried under controlled chemical conditions for characterization and experimentation. The UV-Vis spectrum showed a characteristic peak of AgNPs at 418 nm. FTIR measurements displayed peaks (cm^{-1}) corresponding to various chemical groups [O-H (~3317), C=O (~2102), N-H (~1635), and C-H (~667)]. These peaks indicated the presence and active participation of various botanical compounds in the formation of biogenic NPs. The XRD spectrum exhibited *fcc* structure (size: 36–55 nm) and an average size of ~46 nm. FESEM images showed the dominance of spherically shaped AgNPs with the same sizes (diameters) as calculated from XRD data. Antibacterial activities were measured on five different bacteria: *Staphylococcus aureus, Escherichia coli, Salmonella typhi, Pseudomonas aeruginosa,* and *Vibrio parahemolyticus,* using the well diffusion array technique. Inhibition zones of various sizes, 15–17 mm, with the highest value (17 mm) against *Staphylococcus aureus,* were observed.

William and his colleagues [38] synthesized AgNPs of different sizes using a mixture of leaf extract from the *Rhizophora stylosa* mangrove plant and $AgNO_3$ under various controlled chemical conditions. They observed a gradual color change (pale yellow → dark brown), supporting progressive transformation from $AgNO_3$ → stable spherical colloidal AgNPs. This confirmed that the extract worked as both the reducing/capping agent. The researchers employed several analytical techniques to determine the formation, structure, crystallinity, shape, size, and reduction compounds of these plant-mediated biosynthesized NPs from the leaf extract of *Rhizophora stylosa* mangrove. UV-Vis results displayed three peaks (439 nm, 448 nm, and 453 nm, respectively). These results were achieved by preparing three samples using different concentrations of leaf extract and $AgNO_3$. This is attributed to the fact that the peak positions depend on size and shape, with the peak position shifting from blue to red (i.e., from shorter to longer wavelengths) as the particle size increases. FTIR spectra of the samples showed that diverse compounds of the leaf played vital roles in the formation of stable NPs. The XRD pattern revealed *fcc* crystals of various sizes (25–32 nm). TEM images displayed spherical shapes of the NPs of different sizes (9–32 nm) with an average size of ~30 nm. They further conducted the antibacterial and antioxidant assays of these NPs on various microorganisms and concluded that these biosynthesized AgNPs could potentially create significant inhibition zones against these species.

Cucumis prophetarum is popularly known for its medicinal properties (anticancer, antioxidant, hepatoprotective, and antidiabetic effects) [39]. Hemlata *et al.* [40] published the results of biosynthesized AgNPs from the leaf extract of *Cucumis prophetarum* for its reported medicinal properties. They mixed the leaf extract with $AgNO_3$ to synthesize NPs and found that *Cucumis prophetarum* played the role of both a reducing and capping agent. The formation of NPs was confirmed by a color change of the mixture (pale yellow → reddish-brown) after three hours, signifying the transformation of Ag^+ ions into AgNPs. UV-Vis exhibited a peak at 420 nm, attributed to SPR electrons, further confirming the formation of NPs. Dynamic light scattering measurements provided a particle with an average size of ~90 nm and surface charge potential of ~-37 mV. FTIR results indicated the involvement of various metabolites from *Cucumis prophetarum* in the fabrication of NPs. XRD patterns confirmed the formation of an *fcc* crystalline structure for these green-synthesized AgNPs. SEM images revealed varying shapes (primarily spherical), uniform distribution, and varying sizes (30–50 nm). EDX demonstrated a significant presence of Ag. The researchers assessed the antibacterial behavior of these NPs against gram-positive and gram-negative bacteria, noting distinct inhibition zones for different bacterial types. Additionally, they evaluated the antiproliferative efficacy of these biocompatible AgNPs on various human cancer cells and reported their potential as candidates for cancer treatment.

Kambale *et al.* [41] explored the possibilities of fabricating AgNPs from the leaf extracts of *Congolese* family plants and their antibacterial efficiencies. Three plants from this family, *Brillantaisia patula, Crossopteryx febrifuga*, and *Senna siamea*, were used as bio-resources to synthesize AgNPs and investigate their potential applications in treating microbial-related skin diseases. They cleaned, filtered, and centrifuged the extracts to render them suitable as reducing and capping agents for the biogenic NPs. Subsequently, they mixed the extracts and $AgNO_3$, utilizing suitable parameters for producing the phytochemical-mediated AgNPs. A change in the color of the mixture (yellowish-green → brown) indicated a successful reduction of Ag^+ ions into Ag^0 by the botanic compounds (e.g., polyphenols, alkaloids, and terpenoids) in the leaf extracts, thus confirming the formation of stable colloidal NPs through self-assembly. The green-synthesized samples were further processed for characterization. UV-Vis spectra showed three distinct SPR peaks, characteristic of metallic NPs, at ~430 nm (*Brillantaisia patula*), ~520 nm (*Senna siamea*), and ~490 nm (*Crossopteryx febrifuga*). Variation in the positions of peaks indicates the existence of different chemical constituents in samples since they were prepared from three different plants. TEM images revealed information about the morphologies of three samples. The shapes were found to be spherical in nature, with sizes varying from one species to another: ~45 nm for *Brillantaisia patula*, ~110 nm for *Senna siamea*, and ~45 nm for *Crossopteryx febrifuga*. EDX measurements demonstrated the significant presence of Ag and other chemicals from the extracts in these biogenic NPs. FTIR results demonstrated that the various antibacterial chemical compositions from the leaf extracts played substantial roles as both reduction and capping elements of green synthesized NPs. XRD spectra exhibited the formation of *fcc* crystal structures. Researchers claimed that all measurements complement each other. Moreover, they tested the applicability of the NPs as potential candidates for treating infectious skin diseases. They used three different types of bacteria: *E. coli* and *P. aeruginosa* (gram-negative) and *S. aureus* (gram-positive). They found that *Brillantaisia patula* and *Crossopteryx febrifuga* synthesized AgNPs are more effective on *E. coli* and *P. aeruginosa* than *S. aureus* bacteria. This observation was consistent with previously reported results indicating that Ag^+ ions interact more efficiently with gram-negative bacteria due to their thinner cell layer compared to gram-positive bacteria. However, the opposite trend was observed for *Senna siamea*, likely because the extract of *Senna siamea* has been reported to have several times higher efficacy against *S. aureus* compared to *E. coli* and *P. aeruginosa* bacteria.

Krithiga and his group [42] reported on the biosynthesis, properties, and anti-bacterial efficiency of AgNPs from two different plants. They used leaf extracts of *Clitoria ternatea* and *Solanum nigrum* from the families of Fabaceae (genus: *Clitoria*) and Solanaceae (genus: *Solanum*), respectively, which were mixed with $AgNO_3$ solution under standard biosynthesis protocols at ambient conditions. The temperature, time, and pH were optimized to control the reaction process, as well as the formation and stability of the NPs. Biosynthesized AgNPs exhibited SPR peaks at approximately 420 nm (~440 nm) for *Clitoria ternatea* (*Solanum nigrum*). FTIR results indicated that biomolecules from the leaf extracts served as reducing, capping, and stabilizing agents for the NPs. XRD measurements determined spherical shapes and average particle sizes of approximately 20 nm (28 nm) for *Clitoria ternatea* (*Solanum nigrum*). SEM images demonstrated the uniform distribution of particles. The researchers assessed the antibacterial efficacy of AgNPs on various nosocomial pathogens. They concluded that they have potential as effective deterrents against these microorganisms.

Erdogan *et al.* [43] prepared biofriendly and environmentally sustainable AgNPs using the green chemistry route. The researchers used the leaf extract of *Artichoke* from the *Cynara scolymus* family as a reducing agent. The resulting NPs were analyzed by UV-Vis, FTIR, zeta potential, and SEM to determine SPR, reducing agent, and morphology of the NPs. The researchers further conducted noninvasive photodynamic therapy to generate ROS using these AgNPs and demonstrated the potential of these eco-friendly NPs derived from *Cynara scolymus* leaf extract for cancer treatment.

Punica granatum (pomegranate) (family: *Lythraceae*) is known for its various therapeutic effects for various diseases such as dysentery and diarrhea, owing to the rich presence of numerous botanic compounds (phytochemicals) (e.g., tannins, alkaloids, steroids, terpenoids, and flavonoids) [44].

Kumar *et al.* [45] developed AgNPs by stirring *Punica granatum* leaf extract in $AgNO_3$ for 2 hours at 90°C. The mixture color changed gradually (yellow → dark brown), suggesting the evolution of NPs and confirming the reduction properties of *Punica granatum*. Afterward, the sample was allowed to cool down and subsequently was centrifuged at room temperature. UV-Vis showed peaks at ~255 and ~420 nm, respectively. TEM images displayed that biosynthesized AgNPs have different shapes (spherical, rhomboidal, and cubical) with various diameters (88–120 nm). FTIR measurements revealed that different biomolecules from the extract actively participated in this green chemistry method.

Amooaghaie *et al.* [46] employed both an eco-friendly green synthesis method and a conventional chemical synthesis method to prepare AgNPs and subsequently compared the toxic effects of these biofunctionalized AgNPs on both animals and plants. In the green synthesis, an extract from the leaves of *Nigella sativa* (*N. sativa*) was utilized, while the chemical process involved the use of sodium borohydrate and sodium citrate. $AgNO_3$ served as the source of Ag^+ ions for both methods. Color change confirmed the synthesis of NPs in both cases. Samples were subjected to characterization through various analytical techniques, including UV-Vis, SEM, and FTIR. In the green synthesis method, diverse botanical compounds played significant roles in the formation of stable AgNPs. The study further studied the cytotoxicity and phytotoxicity effects of both types of AgNPs on mice and various plants. The results indicated that the biogenic AgNPs derived from the leaf extract of *N. sativa* exhibited lower toxicity compared to those produced through the chemical method. This observation underscores the superiority of the stable, eco-friendly, and less toxic green synthesis method over the conventional chemical approach.

Plantago lanceolata (*P. lanceolata*) (family: *Plantaginaceae*) is a plant that contains numerous biocompounds, including primary and secondary metabolites, vitamins, minerals, antioxidants, medicinal and nutritious elements, as well as antimicrobial properties [47]. Due to these attributes, *P. lanceolata* has been traditionally used for medical purposes, such as treating malaria, tuberculosis, diarrhea, dysentery, bites, wounds, and burns, among other ailments.

Shah *et al.* [48] employed a controlled chemical approach, mixing the leaf extract of the *P. lanceolata* plant with 1 mM of silver chloride (AgCl) solution at a 1:9 ratio at normal temperature to fabricate NPs. Color change (pale-yellow → reddish-brown) confirmed the formation of biogenic NPs. The resulting mixture was further purified, dried, and characterized. UV-Vis exhibited a peak at ~430 nm, affirming successful synthesis. FTIR revealed peaks at various wavelengths attributed to different functional groups (e.g., C=O and C-H), underscoring their presence and involvement in the reduction process. AFM images depicted homogeneous circular NPs with an average size of ~55 nm. XRD data confirmed an *fcc* crystal structure with an average size of ~30 nm, consistent with SEM and TEM images. EDX analysis highlighted the prevalence of the Ag element. Diffuse reflection spectroscopy (DRS) unveiled a band gap (E_g) of 2.34 eV. Brunauer-Emmett-Teller (BET) measurements using N_2 gas revealed the highest surface area of ~7.4 m^2/g, along with a pore volume of ~0.04 mL/g and a radius of ~1.1 nm. XPS analysis indicated predominant Ag and associated oxides' binding energies in the energy range of 367 eV and 375 eV. Antibacterial efficacies of the green synthesized AgNPs were tested against four different bacteria, *Escherichia coli*, *Agrobacterium tumefaciens*, *Proteus vulgaris*, and *Staphylococcus aureus*, resulting in varying inhibition zones. Additionally, antioxidant properties were evaluated using DPPH, and the results exhibited an increasing trend with concentration. In conclusion, the researchers suggested that *P. lanceolata* holds promise as an effective agent for synthesizing AgNPs with potential applications.

Perilla frutescens (L.) plant (family: *Lamiaceae*) is known for its many medicinal properties [49]. Tavan *et al.* [50] used the extract of leaves from the *Perilla frutescens (L.)* plant (family: *Lamiaceae*) for the biosynthesis of AgNPs. They processed the fresh leaves under controlled chemical conditions to develop solutions of different concentrations, which were then preserved. They used $AgNO_3$ reagent of various concentrations for the synthesis processes. They prepared the NPs by mixing the extract and $AgNO_3$ solution, then purified and stored them under controlled chemical conditions for their characterizations and applications. The gradual change of color (various shades of yellow → brown), based on the concentration of $AgNO_3$ solution, indicated the synthesis of NPs. UV-Vis of biosynthesized NPs exhibited a peak with maximum intensity at ~436 nm in all samples, indicating that biomolecules of extract reduced $Ag^+ \rightarrow Ag^o$ and capped Ag^o ions into stable AgNPs. XRD data demonstrated the formation of an *fcc* lattice with an average size of ~15 nm. FESEM showed a dominance of spherical and oval-shaped NPs with sizes (diameter) smaller than 50 nm. EDX measurements unveiled a strong Ag peak at 3 keV, along with a few minor peaks due to the presence of other elements. FTIR analysis displayed several peaks at different wavelengths (cm^{-1}) (O-H ~3462, C=O ~1640, etc.) such as hydroxyl, carbonyl, and alcohols, clearly indicating the significant presence of a variety of biochemicals of medicinal properties in the AgNPs. Surface-enhanced Raman spectroscopy displayed only a strong characteristic single peak of AgNPs at ~1780 cm^{-1}. Zeta potential analysis of AgNPs produced a value of approximately -18 mV, signifying a negative charge. DLS results closely matched with the FESEM measurement of particle size (diameter). Effects of parameters (e.g., concentration, temperature, time, and pH) on the synthesis process, quality, size, and shape of AgNPs were investigated. Furthermore, the researchers characterized solely the leaf extract of the *P. frutescens* plant and compared the results with AgNPs synthesized from *P. frutescens*. The antioxidant abilities of AgNPs using different concentrations were conducted on DPPH and FRAP (reduction of ferric Fe^{3+} ions into ferrous Fe^{+2} ions) and were found to be more effective than the leaf extract. They conducted antibacterial efficiency of AgNPs on gram-positive *Staphylococcus aureus* and gram-negative *Escherichia coli*, as well as antifungal efficiency on *Candida albicans*. The results indicated that green synthesized AgNPs were more efficient than the leaf extract. Both AgNPs and leaf extract were found to be biocompatible with 3-(4, 5-dimethylthiazol-2-yl)-2, 5-diphenyltetrazolium bromide (MTT), suggesting that they could be used without causing substantial damage to normal cells. Similar results were observed for the cytotoxic properties of both AgNPs and leaf extract on breast cancer cells.

Hossain and his group [51] reported the fabrication of AgNPs utilizing leaf extract from the *Alocasia indica* plant (family: *Araceae*) and their potential dental applications. The leaf extract and $AgNO_3$ solution were mixed at room temperature. A color change, after several hours, indicated the synthesis of NPs, which were further processed to prepare the powder form. UV-Vis spectroscopic analysis was conducted at different temperatures, and the SPR peak was observed at 267 nm, which differed from reported values. The optical band gap (E_g) was calculated to be 4.86 eV, also differing from the reported literature values. They suggested that this could be due to the quantum confinement effect. FTIR measurements exhibited peaks at different wavelengths (cm^{-1}) associated with stretching/bending of different functional groups, e.g., C-Br (678), C=C (974), S=O (1394), and O-H (3284), indicating roles of them in the formation of NPs. SEM displayed the formation of circular NPs with sizes varying from ~14 nm to 45 nm. EDX measurements showed the existence of other elements (C, O, and P) in addition to the dominant Ag peak (~2.98 keV). The antibacterial efficacy of AgNPs on gram-positive *Staphylococcus aureus* was found to be higher than the reported values due to the smaller size. The researchers concluded that these AgNPs could be potentially used for implantation in dental cavities.

Numerous industrial sectors, including textiles, leather, pharmaceuticals, cosmetics, food, and plastics, utilize dyes in their production processes [52]. However, the improper disposal of waste containing dyes can result in environmental contamination, presenting a significant risk to ecosystems. This is due to the fact that dyes, being synthetic organic compounds, often contain a range of hazardous chemicals. Various methods, such as electrolysis, adsorption, oxidation, biochemical processes, biodegradation, filtration, and ozonization, have been employed to eliminate or degrade dyes, aiming for a more sustainable ecosystem [52]. However, these methods often fall short of achieving the desired level of efficacy in contaminant removal. As a result, there is a need for the implementation of more environmentally friendly approaches.

Piper colubrinum plant is known for its resistive properties against pathogens (e.g., *Phytophthora capsici*) [53]. Santhoshkumar *et al.* [54] utilized a straightforward plant-assisted approach to synthesize AgNPs using the *Piper colubrinum* plant. They combined an extract derived from 5 g of *Piper colubrinum* leaves with a 1 mM $AgNO_3$ aqueous solution at a 9:1 ratio under controlled and predetermined conditions. The change in color of the mixture (pale-yellow → dark brown), corresponding to SPR, visibly indicated the involvement of phytochemicals in the reaction kinetics. The catalytic and antibacterial properties of NPs were examined. UV-Vis spectra displayed three distinct peaks at 299, 270, and 410 nm for three samples: $AgNO_3$, leaf extract, and green-synthesized biocompatible AgNPs, respectively. These values were consistent with previously reported data, as the samples were individually studied to confirm their purities. The presence of the SPR peak at 410 nm confirmed the involvement of leaf extract in this green synthesis method. FTIR spectra clearly demonstrated the involvement of phytochemicals in the synthesis of stable AgNPs through the reduction of Ag^+ ions into neutral Ag^0. XRD data revealed *fcc* crystal structure with a lattice constant of ~4 Å. TEM, SEM, EDX, and AFM measurements showed closely packed and homogeneous spherical NPs of different sizes (10–50 nm). The catalytic properties of AgNPs were evaluated with three different dyes: Methylene Blue, Bromophenol Blue, and Coomassie Brilliant Blue. The AgNPs displayed significant catalytic efficiency, indicating their potential use in dye reduction for environmental sustainability. Antibacterial efficacy was further investigated using the disc diffusion technique on *Escherichia coli* and *Staphylococcus aureus*. The results revealed significant inhibition zones with a diameter of approximately 30 nm for both species. This observation further corroborated the potent antibacterial properties of these environmentally friendly synthesized AgNPs, consistent with findings reported in other literature.

A few more relevant works are given below (Table 6.1):

Table 6.1 Characterizations and applications of leaf extract-assisted AgNPs.

Leaf extract as reducing agent	Characterizations	Activities/ applications	Ref. no.
Urtica dioica	UV-Vis, XRD, FTIR, Zeta-potential, SEM, EDX, TEM	Antibacterial.	[55]
Melissa officinalis	TEM, XRD, UV-Vis, Raman, FTIR	Antibacterial.	[56]
Parkia speciosa	UV–Vis, SEM, TEM, XRD, EDX, DLS.	Photocatalytic, antibacterial, and antioxidant.	[57]
Punica granatum, Cydonia oblonga, Castanea sativa, Ficus carica, Juglans cinerea, Morus nigra, and *Morus alba*	UV-Vis, TEM.	Antioxidant.	[58]
Azadirachta indica	UV-Vis, TEM, FTIR, DLS.	Antibacterial.	[59]
Azadirachta indica	XRD, SEM, FTIR, PL	Antibacterial.	[60]
Carya illinoinensis	UV-Vis, XRD, TEM, SEM.	Antibacterial.	[61]

6.3 Conclusions

In summary, we have discussed the cost-effective, nontoxic, fast, and eco-friendly green synthesis of biocompatible AgNPs using leaf extracts of various medicinal plants. We explored their characterizations through various analytical spectroscopic techniques and their applications in the field of nanobiotechnology/nanobiomedicine and environmental sustainability. Various biomolecules associated with leaf extracts played significant roles in reducing the silver ions (Ag^+) from $AgNO_3$ to neutral Ag (Ag^o) and in capping and forming biogenic AgNPs. The color change of the mixture of leaf extract and $AgNO_3$ reagent confirmed AgNPs' formation. UV-Vis results displayed characteristic peaks of AgNPs due to SPR at specific wavelengths, further confirming their formation. FTIR measurements exhibited the presence of various phytochemicals/functional groups, such as O-H (phenols), C=O (carboxylic acids), N-H (amines), and C-H (alkynes), in the leaf extract, indicating their active participation in AgNPs' synthesis. XRD data revealed the formation of *fcc* crystal structure in most cases, albeit with varying sizes. TEM, HRTEM, SEM, and AFM images demonstrated mostly spherical shapes with varying dispersion and sizes. AgNPs demonstrated effective antimicrobial efficacy against various types of bacteria/viruses, many of which cause life-threatening diseases. This underscores the promising biocidal properties of biosynthesized AgNPs in the field of nanomedicine. Furthermore, AgNPs exhibited catalytic properties on dyes, implying their potential use in mitigating environmental pollution. However, a further systematic review of leaf extract-assisted green synthesis of AgNPs is necessary to comprehensively understand this emerging field.

References

1. Harish, K.K., Venkatesh, N., Bhowmik, H., Kuila, A., Metallic Nanoparticle: A Review. *Biomed. J. Sci. Tech. Res.*, 4, 2, 3765–3775, 2018.
2. Sowmya, N., Mohanty, D., Nirosha, B., Uttam Kumar, N., Kumar Patra, P., A Systematic Review of Metallic Nanoparticles: Synthesis, Biological Activities & Applications. *J. Pharm. Negat. Results*, 14, Special Issue 2, 2525–2533, 2023.
3. Alaqad, K. and Saleh, T.A., Gold and Silver Nanoparticles: Synthesis Methods, Characterization Routes and Applications towards Drugs. *J. Environ. Anal. Toxicol.*, 6, 4, 1000384, 2016.
4. Jeyaraj, M., Gurunathan, S., Qasim, M., Kang, M.-H., Kim, J.-H., A Comprehensive Review on the Synthesis, Characterization, and Biomedical Application of Platinum Nanoparticles. *Nanomaterials*, 9, 1719, 2019.
5. Bhagat, M., Anand, R., Sharma, P., Rajput, P., Sharma, N., Singh, K., Review—Multifunctional Copper Nanoparticles: Synthesis and Applications. *ECS J. Solid State Sci. Technol.*, 10, 063011, 2021.

6. Patil, S. and Chandrasekaran, R., Biogenic nanoparticles: a comprehensive perspective in synthesis, characterization, application and its challenges. *J. Genet. Eng. Biotechnol.*, 18, 67, 2020.
7. Jaiswal, S., Roy, N., Dhwaj, A., Verma, D., Prabhakar, A., A Review on the Various Mechanisms of Green Synthesis of Metal Nanoparticles for Biomedical Applications. *Curr. Green Synth.*, 9, 2, 62–73, 2022.
8. Adeyemi, J.O., Oriola, A.O., Onwudiwe, D.C., Oyedeji, A.O., Plant Extracts Mediated Metal-Based Nanoparticles: Synthesis and Biological Applications. *Biomolecules*, 12, 627, 2022.
9. Ribeiro, A.I., Dias, A.M., Zille, A., Synergistic Effects Between Metal Nanoparticles and Commercial Antimicrobial Agents: A Review. *ACS Appl. Nano Mater.*, 5, 3030–3064, 2022.
10. Slavin, Y.N., Asnis, J., Häfeli, U.O., Bach, H., Metal nanoparticles: understanding the mechanisms behind antibacterial activity. *J. Nanobiotechnol.*, 15, 65, 1–20, 2017.
11. Kulkarni, N. and Muddapur, U., Biosynthesis of Metal Nanoparticles: A Review. *J. Nanotechnol.*, 2014, Article ID 510246, 8, 2014.
12. Agrawal, S., Bhatt, M., Kumar Rai, S., Bhatt, A., Dangwal, P., Kumar Agrawal, P., Silver nanoparticles and its potential applications: A review. *J. Pharmacogn. Phytochem.*, 7, 2, 930–937, 2018.
13. Zhang, X.-F., Liu, Z.-G., Shen, W., Gurunathan, S., Silver Nanoparticles: Synthesis, Characterization, Properties, Applications, and Therapeutic Approaches. *Int. J. Mol. Sci.*, 17, 1534, 2016.
14. Ahmed, S., Ahmad, M., Swami, B.L., Ikram, S., A review on plants extract mediated synthesis of silver nanoparticles for antimicrobial applications: A green expertise. *J. Adv. Res.*, 7, 17–28, 2016.
15. Vishwanath, R. and Negi, B., Conventional and green methods of synthesis of silver nanoparticles and their antimicrobial properties. *Curr. Res. Green Sustain. Chem.*, 4, 100205, 2021.
16. Couto, C. and Almeida, A., Metallic Nanoparticles in the Food Sector: A Mini-Review. *Foods*, 11, 402, 2022.
17. Hun Lee, S. and Jun, B.-H., Silver Nanoparticles: Synthesis and Application for Nanomedicine. *Int. J. Mol. Sci.*, 20, 865, 2019.
18. Gajbhiye, S. and Sakharwade, S., Silver Nanoparticles in Cosmetics. *J. Cosmetics Dermatological Sci. Appl.*, 6, 48–53, 2016.
19. Siddiqi, K.S., Husen, A., Rao, R.A.K., A review on biosynthesis of silver nanoparticles and their biocidal properties. *J. Nanobiotechnol.*, 16, 14, 2018.
20. Vanlalveni, C., Lallianrawna, S., Biswas, A., Selvaraj, M., Changmai, B., Rokhum, S.L., Green synthesis of silver nanoparticles using plant extracts and their antimicrobial activities: a review of recent literature. *RSC Adv.*, 11, 2804, 2021.
21. Chandraker, S.K., Ghosh, M.K., Lal, M., Shukla, R., A review on plant-mediated synthesis of silver nanoparticles, their characterization and applications. *Nano Express*, 2, 022008, 2021.

22. Chung, I.-M., Park, I., Seung-Hyun, K., Thiruvengadam, M., G.R., Plant-Mediated Synthesis of Silver Nanoparticles: Their Characteristic Properties and Therapeutic Applications. *Nanoscale Res. Lett.*, 11, 40, 2016.
23. Ying, S., Guan, Z., Ofoegbu, P.C., Clubb, P., Rico, C., Hea, F., Hong, J., Green synthesis of nanoparticles: Current developments and limitations. *Environ. Technol. Innov.*, 26, 102336, 2022.
24. Sulaiman, C.T. and Balachandran, I., Total phenolics and total flavonoids in selected Indian medicinal plants. *Indian J. Pharm. Sci.*, 74, 3, 258–260, 2012.
25. M.S.S., Chagas, M.D., Behrens, Moragas-Tellis, C.J., Penedo, G.X.M., Silva, A.R., C.F.G.--A., Flavonols and Flavones as Potential anti-Inflammatory, Antioxidant, and Antibacterial Compounds. *Oxid. Med. Cell. Longevity*, 2022, Article ID 9966750, 1–25, 2022.
26. Mutha, R.E., Tatiya, A.U., Surana, S.J., Flavonoids as natural phenolic compounds and their role in therapeutics: an overview. *Future J. Pharm. Sci.*, 7, 25, 2021.
27. Jain, A.S., Pawar, P.S., Sarkar, A., Junnuthula, V., Dyawanapelly, S., Bionanofactories for Green Synthesis of Silver Nanoparticles: Toward Antimicrobial Applications. *Int. J. Mol. Sci.*, 22, 11993, 2021.
28. Anandalakshmi, K., Venugobal, J., Ramasamy, V., Characterization of silver nanoparticles by green synthesis method using Pedalium murex leaf extract and their antibacterial activity. *Appl. Nanosci.*, 6, 399–408, 2016.
29. Ansar, S., Tabassum, H., Aladwan, N.S.M., Ali, M.N., Almaarik, B., AlMahrouqi, S., Abudawood, M., Banu, N., Alsubki, R., Eco friendly silver nanoparticles synthesis by Brassica oleracea and its antibacterial, anticancer and antioxidant properties. *Sci. Rep.*, 10, 18564, 2020.
30. Tamileswari, R., Haniff Nisha, M., Jesurani, S., Green Synthesis of Silver Nanoparticles using Brassica Oleracea (Cauliflower) and Brassica Oleracea Capitata (Cabbage) and the Analysis of Antimicrobial Activity. *Int. J. Eng. Res. Technol.*, 4, 1071, 2015.
31. Singh, A., Sharma, B., Deswal, R., Green silver nanoparticles from novel Brassicaceae cultivars with enhanced antimicrobial potential than earlier reported Brassicaceae members. *J. Trace Elem. Med. Biol.*, 47, 1–11, 2018.
32. Kumar, S., Singh, B., Singh, R., Catharanthus roseus (L.) G. Don: A review of its ethnobotany, phytochemistry, ethnopharmacology and toxicities. *J. Ethnopharmacol.*, 284, 114647, 2022.
33. Ponarulselvam, S., Panneerselvam, C., Murugan, K., Aarthi, N., Kalimuthu, K., Thangamani, S., Synthesis of silver nanoparticles using leaves of Catharanthus roseus Linn. G. Don and their antiplasmodial activities. *Asian Pac. J. Trop. Biomed.*, 2, 7, 574–580, 2012.
34. Alkammash, N.M., Synthesis of Silver Nanoparticles from Artemisia Sieberi and Calotropis Procera Medical Plant Extracts and Their Characterization using SEM Analysis. *Biosci. Biotechnol. Res. Asia*, 14, 2, 521–526, 2017.

35. Ochieng, M.A., Bakrim, W.B., Bitchagno, G.T.M., Mahmoud, M.F., Sobeh, *Syzygium jambos* L. Alston: An Insight Into its Phytochemistry, Traditional Uses, and Pharmacological Properties. M., *Front. Pharmacol.*, 13, 786712, 2022.
36. Dutta, P.P., Bordoloi, M., Gogoi, K., Roy, S., Narzary, B., Bhattacharyya, D.R., Mohapatra, P.K., Mazumder, B., Antimalarial silver and gold nanoparticles: Green synthesis, characterization and *in vitro* study. *Biomed. Pharmacother.*, 91, 567–580, 2017.
37. Thampi, N. and Jeyadoss, V.S., Biogenic Synthesis and Characterization of Silver Nanoparticles Using Syzygium samarangense (Wax Apple) Leaves Extract and Their Antibacterial Activity. *Int. J. PharmTech. Res.*, 8, 3, 426–433, 2015.
38. Willian, N., Syukri, Z., Labanni, A., Arief, S., Bio-Friendly Synthesis of Silver Nanoparticles Using Mangrove Rhizophora stylosa Leaf Aqueous Extract and Its Antibacterial and Antioxidant Activity. *Rasayan J. Chem.*, 13, 3, 1478–1485, 2020.
39. Kavishankar, G. and Lakshmidevi, N., Anti-diabetic effect of a novel N-Trisaccharide isolated from Cucumis prophetarum on streptozotocin-nicotinamide induced type 2 diabetic rats. *Phytomedicine*, 21, 624–630, 2014.
40. Hemlata, P.R.M., Singh, A.P., Tejavath, K.K., Biosynthesis of Silver Nanoparticles Using Cucumis prophetarum Aqueous Leaf Extract and Their Antibacterial and Antiproliferative Activity Against Cancer Cell Lines. *ACS Omega*, 5, 5520–5528, 2020.
41. Kambale, E.K., Nkanga, C., Mutonkole, B.-P., Bapolisi, A.M., Tassa, D.O., Liesse, J.-M., Krause, R.WM., Memvanga, P.B., Green synthesis of antimicrobial silver nanoparticles using aqueous leaf extracts from three Congolese plant species (Brillantaisia patula, Crossopteryx febrifuga and Senna siamea). *Heliyon*, 6, e04493, 2020.
42. Krithiga, N., Rajalakshmi, A., Jayachitra, A., Green Synthesis of Silver Nanoparticles Using Leaf Extracts of Clitoria ternatea and Solanum nigrum and Study of Its Antibacterial Effect against Common Nosocomial Pathogens. *J. Nanosci.*, Article ID 928204, 2015, 8, 2015.
43. Erdogan, O., Abbak, M., Demirbolat, G.M., Birtekocak, F., Aksel, M., Pasa, S., Cevik, O., Green synthesis of silver nanoparticles via Cynara scolymus leaf extracts: The characterization, anticancer potential with photodynamic therapy in MCF7 cells. *PLoS One*, 14, 6, e0216496, 2019.
44. Rahimi, H.R., Arastoo, M., Ostad, S.N., A Comprehensive Review of Punica granatum (Pomegranate) Properties in Toxicological, Pharmacological, Cellular and Molecular Biology Researches. *Iran. J. Pharm. Res.*, 11, 2, 385–400, 2012.
45. Kumar, M., Dandapat, S., Ranjan, R., Kumar, A., Sinha, M.P., Plant mediated synthesis of silver nanoparticles using Punica granatum aqueous leaf extract. *J. Microbiol. Exp.*, 6, 4, 175–178, 2018.

46. Amooaghaie, R., Reza Saeri, M., Azizi, M., Synthesis, characterization and biocompatibility of silver nanoparticles synthesized from Nigella sativa leaf extract in comparison with chemical silver nanoparticles. *Ecotoxicol. Environ. Saf.*, 120, 400–408, 2015.
47. Bahadori, M.B., Sarikurkcu, C., Sefa Kocak, M., Calapoglu, M., Cemil Uren, M., Ceylan, O., Plantago lanceolata as a source of health-beneficial phytochemicals: Phenolics profile and antioxidant capacity. *Food Biosci.*, 34, 100536, 2020.
48. Zahir Shah, M., Guan, ZH., Ud Din, A., Ali, A., Ur Rehman, A., Jan, K., Faisal, S., Saud, S., Adnan, M., Wahid, F., Alamri, S., Siddiqui, M.H., Ali, S., Nasim, W., Hammad, H.M., Fahad, S., Synthesis of silver nanoparticles using Plantago lanceolata extract and assessing their antibacterial and antioxidant activities. *Sci. Rep.*, 11, 20754, 2021.
49. Dhyani, A., Chopra, R., Garg, M., A Review on Nutritional Value, Functional Properties and Pharmacological Application of Perilla (Perilla Frutescens L.). *Biomed Pharmacol. J.*, 12, 2, 649–660, 2019.
50. Tavan, M., Hanachi, P., Mirjalili, M.H., DashtbaniRoozbehani, A., Comparative assessment of the biological activity of the green synthesized silver nanoparticles and aqueous leaf extract of Perilla frutescens (L.). *Sci. Rep.*, 13, 6391, 2023.
51. Hossain, N., Chowdhury, D.M.A., Hossain, A., Ahmed, S., Rana, M., Sultana, Synthesis and Characterization of Alocasia indica Infused Silver Nanoparticles for Dental Implant Applications. S., *Chem. Phys. Impact*, 6, 100239, 2023.
52. Al-Tohamy, R., Ali, S.S., Li, F., Okasha, K.M., Mahmoud, Y.A.-G., Elsamahy, T., Jiao, H., Fu, Y., Sun, J., A critical review on the treatment of dye-containing wastewater: Ecotoxicological and health concerns of textile dyes and possible remediation approaches for environmental safety. *Ecotoxicol. Environ. Saf.*, 231, 113160, 2022.
53. Malik, N. and George, J.K., Resistance Genes in Piper colubrinum: In Silico Survey From Leaf Transcriptome and Expression Studies Upon Challenge Inoculation with Phytophthora capsica. *Appl. Biochem. Biotechnol.*, 184, 987–1008, 2018.
54. Santhoshkumar, R., Hima Parvathy, A., Soniya, E.V., Phytosynthesis of silver nanoparticles from aqueous leaf extracts of Piper colubrinum: characterisation and catalytic activity. *J. Exp. Nanosci.*, 16, 1, 294–308, 2021.
55. Jyoti, K., Baunthiyal, M., Singh, A., Characterization of silver nanoparticles synthesized using Urtica dioica Linn. leaves and their synergistic effects with antibiotics. *J. Rad. Res. Appl. Sci.*, 9, 217–227, 2016.
56. de Jesús Ruíz-Baltazar, A., Reyes-López, S.Y., Larrañaga, D., Estévez, M., Pérez, R., Green synthesis of silver nanoparticles using a Melissa officinalis leaf extract with antibacterial properties. *Results Phys.*, 7, 2639–2643, 2017.

57. Ravichandran, V., Vasanthi, S., Shalini, S., Tripathy, S.A.A.S,.M., Paliwal, N., Green synthesis, characterization, antibacterial, antioxidant and photocatalytic activity of Parkia speciosa leaves extract mediated silver nanoparticles. *Results Phys.*, 15, 102565, 2019.
58. Azat Akbal, M., H.T., A.C., B.U., Relation between Silver Nanoparticle Formation Rate and Antioxidant Capacity of Aqueous Plant Leaf Extracts. *J. Spectro.*, Article ID 4083421, 2016, 6, 2016.
59. Shakeel, A., Saifullah, M.A., Swami, B.L., Ikram, S., Green synthesis of silver nanoparticles using Azadirachta indica aqueous leaf extract. *J. Rad. Res. Appl. Sci.*, 9, 1–7, 2016.
60. Verma, A. and Mehata, M.S., Controllable synthesis of silver nanoparticles using Neem leaves and their antimicrobial activity. *J. Rad. Res. Appl. Sci.*, 9, 109–115, 2016.
61. Bakht Dalir, S.J., Djahaniani, H., Nabati, F., Hekmati, M., Characterization and the evaluation of antimicrobial activities of silver nanoparticles biosynthesized from Carya illinoinensis leaf extract. *Heliyon*, 6, e03624, 2020.

7

Material for Energy Storage and Conversion

Ruma Das[1]* and Abhirup Paria[2]

[1]*Department of Applied Science (Physics), Haldia Institute of Technology, Haldia, India*
[2]*Department of Computer Science & Engineering (Artificial Intelligence & Machine Learning), Haldia Institute of Technology, Haldia, India*

Abstract
To address the ever-growing energy need, the expansion of a highly proficient energy storage and conversion system is inevitably essential. The promising technologies and up-to-the-mark performance of the energy devices are intimately dependent on the electrode materials and structural features of the devices. Two-dimensional (2D) materials and their hybrid structures offer promising growth in energy applications. There are hundreds of layered materials in semiconductors (MoS_2, WS_2), semimetals (graphene, $TiSe_2$, $NiTe_2$), insulators (boron nitride), and superconductors (NbS_2, $NbSe_2$) groups, which are exfoliated to prepare the 2D structure for the energy specific applications. The graphene sheet is the most promising 2D material in this field. After graphitic materials, the chalcogenides, including transition metal dichalcogenide (TMD), have recently been getting attention as 2D materials. Besides inorganic nanostructures with 2D form, organic and organometallic nanosheets with porous structures are emphasized for energy devices. Compared to a bulk structure, 2D nanosheets offer outstanding electronic and optical properties due to high surface exposure, active sites, defect states, tunable bandgap, and flexibility towards heterostructure formation. A few essential factors, such as high electrical conductance, porosity, shorter electrical transportation length, and ion diffusion pathway, are extensively involved in improving the energy storage or conversion system. Large plane areas and a huge count of active sites of 2D nanostructure provide improved mobility of the carriers, high conductivity, and catalytic activity, which support 2D nanosheets for electrochemical

**Corresponding author*: ruma.dasparia@hithaldia.ac.in

Kolla Bhanu Prakash, S. V. Ranganayakulu and K. S. Jagannatha Rao (eds.) New Frontiers in Materials Science: Interdisciplinary Approaches to Innovation and Technologies, (119–140) © 2025 Scrivener Publishing LLC

catalysis and photocatalysis applications. Extraordinary surface area and improved conductivity play a pivotal role in energy storage. In this present study, we focus on various energy storage and conversion systems with several 2D nanostructures, considering supercapacitors and batteries as energy storage systems and oxygen reduction reaction (ORR), oxygen evolution reaction (OER), hydrogen evolution reaction (HER), carbon dioxide reduction reactions (CRR), and water splitting as the energy conversion process.

Keywords: 2D nanomaterials, hybrid structures, renewable energy storage (battery, supercapacitor), energy conversion

7.1 Introduction

With the severe upsurge in energy demands and environmental pollution, the growth of advanced energy machinery is essential to utilize renewable energy sources in the best possible ways. The utilization of renewable energy offers an ecological alternative to avoid fossil fuels and related environmental concerns. For the assistance of energy technology, the growth of advanced and promising materials is highly essential. In the present date, two-dimensional (2D) nanomaterials owing to their exceptional behavior, have appeared as encouraging entrants for energy technology [1–3]. To design the electrodes and catalysts, the tunable bandgaps and ability of the hybrid structure formation of 2D nanomaterials make their keen interest in energy technology. There are thousands of layered structure materials found in nature, including insulators (hBN, HfS_2), semiconductors (MoS_2, WS_2, $MoSe_2$, TaS_2, etc.), semimetals (graphene, $NiTe_2$, WTe_2, etc.) and superconductors (NbS_2, $TaSe_2$, etc.), which are exfoliated to synthesize 2D nanosheets with extraordinary physical and chemical features [4–6]. Graphene is one of the materials that is growing exponentially in the 2D nanomaterial world. Along with graphene, its derivative materials like graphene oxide (GO), reduced GO (RGO), doped and hybrid graphene, graphdiyne, and graphane are also getting extensive attention in this nanomaterial field [7–9]. Other emerging 2D nanosheets are the chalcogenides groups which include transition metal dichalcogenides (TMDs), metal trichalcogenides (Bi_2Te_3, $TaSe_3$, $NbSe_3$), as well as GaSe, InSe, SnS_2, and SnS [10, 11]. MoS_2 is a well-established TMD material, and other TMD materials with MX_2 structure (M: transition metals like Ti, Zr, and X: Te, S, Se) are recently getting attention [10, 12, 13]. Presently, a group of crystalline metals (Au, Pd, Pt, Ni, Ru) and metal oxides (TiO_2, Fe_2O_3, Al_2O_3, Ga_2O_3, WO_3, $BiVO_4$, In_2O_3, etc.) along with chalcogenides (ZnS, CuSe, ZnSe, CdSe, $CoSe_2$, etc.) are observed to be reported in 2D material class [14]. Along with inorganic 2D nanostructure, organic and organometallic nanomaterials are drawing attention with their unique porous framework comprised of covalent organic and metal-organic structures and polymer, which are able to be modified at the molecular level by changing the ligands and linkages [15, 16]. The existence of a huge number of 2D materials encourages the preparation of various hybrid structures with remarkable properties, particularly for energy applications. Compared to bulk materials, there is a modification of the bandgap and, in general, an enhancement of the carrier concentrations and electron mobility observed in nanosheets. Due to the 2D structure, the specific surface area increases along with a greater exposure of the surface, enlarging the availability of active states for photocatalysis or redox reactions process. The active states also facilitate the storage of more ions in energy storage devices. The transportation of ions is highly improved by the 2D structure of electrode materials. Moreover, the upsurge conductivity of nanosheets facilitates the high power density in energy storage. The present vacancies in nanosheets actively participate in the catalysis process by trapping large numbers of carriers. Enormous edge sites and functional groups in 2D sheets also affect the catalytic applications.

This chapter emphasizes the advancement of 2D nanosheets for transforming and storing renewable energy. The presence of numerous 2D nanostructures in nature with their unique properties towards renewable energy sources is highlighted. Metal-free 2D structures are the recent trend for low-cost, efficient catalysis and electrode materials, which are also discussed here. Along with these, the implementation of hybrid nanosheets is light up for the enhancement of energy devices, particularly for supercapacitors and batteries as energy storage systems and ORR, OER, HER, CRR, and water splitting as energy conversion processes. After that, the challenges and the scope of the 2D nanomaterials and their hybrid structures are pointed out.

7.2 2D Materials in Energy Application

To meet the increasing energy crisis, 2D nanomaterials are extensively implemented for the improvement of energy storage and adaptation technology. Figure 7.1 shows data visualization of 2D nanosheet material-based energy conversion and storage work in the period of 2014–2024. The image appears to be a VOSviewer visualization of a bibliometric analysis, focusing on research related to energy storage and conversion. The network diagram represents the interconnection between various terms and concepts within this field. The terms 'energy conversion,' 'energy storage,' and 'electrodes' occupy central positions in the network, suggesting their

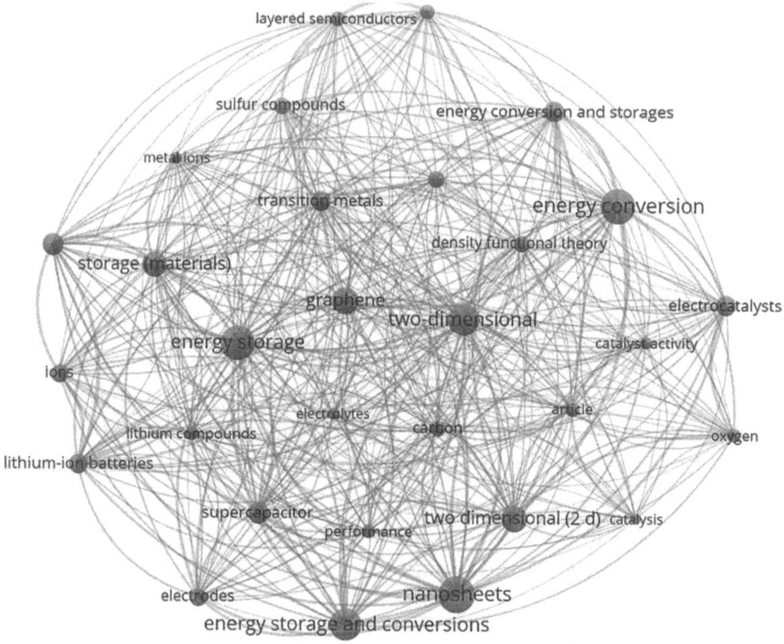

Figure 7.1 Data visualization of 2D nanosheet material-based energy conversion and storage work in the period of 2014–2024.

importance as core topics within the research area. The main terms such as layered semiconductors, sulfur compounds, transition metals, density functional theory, graphene, two-dimensional electrocatalysts, ions, lithium compounds, electrolytes, carbon, lithium-ion batteries, supercapacitor, performance, nanosheets, and energy storage and conversions are clustered around these central themes, indicating their strong association with the broader field. The network exhibits sub-clusters or communities of terms, suggesting the existence of distinct research areas within the energy storage and conversion domain. Interestingly, the terms related to lithium-ion batteries and supercapacitors form separate groups. The lines connecting the terms represent the relationships between them in the analyzed literature. The thickness of the lines corresponds to the frequency or strength of such relationships. Generally, the visualization provides a visual representation of the knowledge structure and interconnections within the field of energy storage and conversion, highlighting the key concepts, relationships, and potential research areas. It explores the existing literature, identifies emerging trends, and identifies potential gaps in knowledge.

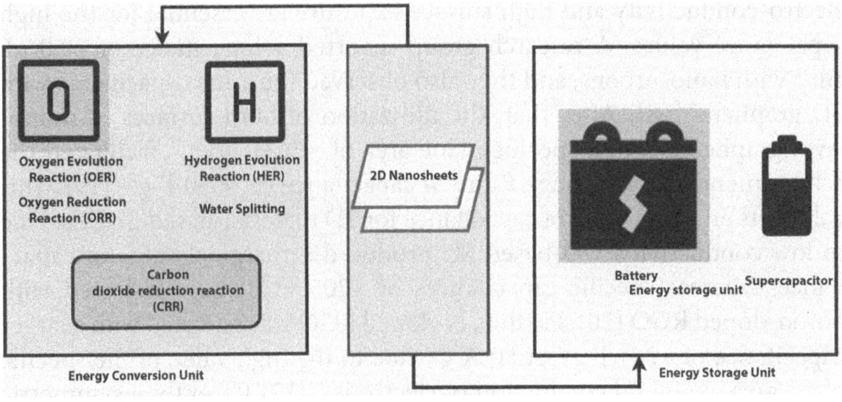

Figure 7.2 Schematic of the green energy applications of 2D nanosheets.

Nanosheets with large surface exposure, abundant active sites, enormous functional groups, vacancies, dangling bonds, and high porosity direct researchers for their implementation in various energy applications. Here, we will discuss the use of various 2D nanomaterials as energy storage applications (supercapacitor and battery) and energy conversion applications (O_2 and CO_2 reduction reaction, oxygen and H_2 evolution reaction, and water splitting). A schematic of the green energy applications of 2D nanosheets is presented in Figure 7.2.

7.2.1 Materials Used in Energy Storage

7.2.1.1 Supercapacitors

Supercapacitors (SC), also known as ultracapacitors or electrochemical capacitors, are storage devices with high control density. The capacitive value and the density of power of SCs are much higher than those of conventional capacitors. Since 2009, SCs have been commercialized for transportation, particularly for city transit buses, trams, and automobiles [17]. SCs are mainly of two types: electrical dual-layer capacitors (EDLC) and pseudocapacitors (PSCs). In EDLC, an electrostatic charge is built up via electrostatic interaction at the frontier of the electrodes and the electrolyte using a non-Faradaic effect. In the case of PSC, the energy is stored following the same process as EDLCs, with the addition of Faradaic redox reactions on the electrode's surface. PSCs are basically intermediate devices between EDLCs and batteries. For EDLCs, a dielectric medium with ion diffusion and migration ability and electrode materials with high electro-conductivity and high surface exposure are essential for the high capacitance values. A research group reported a capacitance of 20.0 μF cm^{-2} with nanocarbons, and they also observed the same capacitance with 2D graphene [18], After that, the utilization of both surfaces of mono-layer graphene, with a specific plane area of ~2631 m^2 g^{-1}, facilitated the achievement of a very huge value of capacitance of ~550 F g^{-1} [19]. This value sets an upper limit of capacitance for 2D material-based EDLCs. Due to low conductivity, GO-based SC produced a marginal value of capacitance, whereas specific capacitances of ~205 F g^{-1} were reported with boron-doped RGO [20]. Further, N-doped RGO was reported with a larger capacitance of ~261 F g^{-1} at 50 A g^{-1} due to the high value of the specific plane area and good conduction of electrodes. [17] Recently, a symmetric SC device has been reported with activated RGO-carbon cloth composite electrodes, which attained a capacitance of 380.2 mF cm^{-2} at 5 mV/s [21]. Additionally, a comparative study of symmetric and asymmetric electrodes in SC (activated RGO sheet wrapped carbon as anode and MXene-coated carbon as cathode) increased the potential range of ~1 V–1.6 V and the highest energy density of ~107 μWh cm^{-2} at a power density of 485 μW cm^{-2}, which is nearly two times greater than symmetric SC. Noteworthy, the asymmetric SC device displays ~90% retention capacity after repeated twisting for more than 5000 times and 10000 cycles of charging and discharging. With a flexible polyethylene terephthalate (PET) substrate, RGO electrode coated with H_2SO_4/PVA solid-state electrolyte was used in EDLC with a density of power of ~200 W cm^{-3}. Noteworthy, a high value of the power density and energy storage remains unchanged even after severe bending of the substrate, indicating the high applicability of this EDLC in wearable and printed electronic devices. They also demonstrated that the use of ionogel as an electrolyte upsurges the potential in the span of 1 V–2.5 V, along with an enhancement of the specific capacitance. From the MXene family, titanium carbides ($Ti_3C_2T_x$, T_x: oxygen/halide/hydroxide) can be cast as electrodes for EDLC due to their huge electrical conductivity and polarization. $Ti_3C_2T_x$, prepared by etching of Ti_3AlC_2, intercalated numerous cations within a multi-layered nanosheet, and that helped the creation of an electric layer with charge in EDLCs. Levi *et al.* stated a capacitance value of 200 F g^{-1} in KOH solution with 2D $Ti_3C_2T_x$ [22]. Zhang *et al.* used $Ti_3C_2T_x$ as transparent and flexible electrodes with charge/discharge cycles of more than ~20,000 [23]. In the TMD group, MoS_2 and VS_2 were studied for flexible electronic devices. A capacitance of 100.0 F g^{-1} was achieved with exfoliated MoS_2 nanosheets, which were exfoliated in an aqueous salt medium [24]. This medium incorporated a negative charge at the MoS_2 nanosheet surfaces, and this facilitated the intercalation of alkali cations within the nanostructure, yielding a high value of capacitance. VS_2 was demonstrated as the flexible thin-film electrode by coating cellulose membranes for flexible and hard devices.

Compared to EDLCs, PSCs offer greater energy storage density, though the power density is still in the developing stage due to the constraint of the conductivity of the layers [7]. In PSCs, conducting EDLC nanosheets are generally used as the conducting material upon which novel inorganic nanosheets are mounted for efficient absorption of electrons, intercalation of ions, and redox reactions. The electrodes of PSCs with inorganic nanosheets hybrid materials and graphene nanosheets have recently been getting interest in research materials. Recently, layered double hydroxides (LDH) materials hybrid with RGO were reported with capacitance values of ~3610 and 3344 F g^{-1} at 2 and 1.25 A g^{-1}, respectively. [25, 26] The hybrid structure of RGO and $Ni(OH)_2/MnO_2$ formed capacitance values of ~2627 and 1327 F g^{-1} at 3 and 20 A g^{-1}, respectively [27]. PSCs offer larger energy storage capacity compared to EDLCs and establish an intermediate technology between classical capacitors and batteries. Other hybrid electrodes like Co_3O_4/Ti, Ni_3S_2/Ni foam, $(NiCo)_3BO_3/Ni(OH)_2$/Ni foam are used as the efficient electrode of PSCs.

7.2.1.2 Battery

The rechargeable battery is a potential storage device that accumulates chemical energy and later converts it to electrical energy through periodic charging and discharging cycles. Li-ion batteries (LIBs), Na-ion batteries (SIBs), and other developing batteries such as Mg-ion, metal-air, and metal-sulfur batteries have been in the category of rechargeable batteries that have recently been studied. Compared with outmoded nickel-cadmium and lead acid batteries, LIBs are much more compact and lightweight, with greater energy density and a larger number of life cycles. Still, the low availability, high cost, and safety issues are prohibiting the large-scale implementation of LIBs. On the other hand, SIBs are getting advantages owing to the low cost of sodium, high availability in nature, and high value of theoretical capacity. However, the bigger size of the radius of Na^+ ion (~1.02 Å) and its lower reduction potential produce lower transportation of Na^+ ions, along with low gravimetric and volumetric energy densities. Compared to quantum dots or nanorods, 2D nanostructures produce larger specific capacities due to high surface areas with enormous active sites facilitating higher storage of cations. Moreover, the atomic-level thick layers yield high mechanical elasticity, presenting them for the development of stretchable, foldable, and thin devices. Certainly, 2D nanostructures establish themselves as promising materials in the field of rechargeable batteries with superior skill and larger cycling consistency compared to their bulk counterparts. Recently, graphene and graphene-associated materials have been explored in the technology of rechargeable batteries. The implementation of graphene as electrodes (cathode, anode, or both) allows rapid transportation of ions and electrons, allowing fast charging and discharging. Dudney *et al.* reported graphene-based LIBs using graphene as an anode, where Li^+ ions were attached to both surfaces of graphene by the formation of Li_2C_6, and this LIB has a theoretical capacitance value of ~744 mAh g^{-1} [28]. RGO was implemented in LIBs with an early discharge capacitance of 540–1600 mAh g^{-1} at 50 mA g^{-1}, depending on the surface-to-volume ratio, more interplane spacing, chemical functionalities, morphology, and active sites [29]. The defect states and edge sites offer lithiation/delithiation reactions and help for Li^+ storage, providing a greater specific capacity. Doping graphene with N, S, B, F, and P modifies the atomic charge concentration, spin concentration, and interplanar spacing and can produce additional uncovered active states, which cause fast transportation of Li^+ ions by absorption-diffusion process on the electrode, facilitating high rate transportation of electrons and ions enhancing entire battery capacities to more than ~1000 mAh g^{-1} at 50 mA g^{-1} [30]. Doping heteroatoms N, S, and P into graphene structure increases electron density in graphene while doping with B creates a deficiency in the graphene lattice. Noteworthy, this property of B-doped graphene offers a smaller potential barrier in the diffusion of Li^+ compared to other doped graphene [31]. Hardikar *et al.* theoretically calculated a very large value of specific capacitance of ~2271 mAh g^{-1} for Li storage with B-doped graphene as the anode [32]. Das *et al.* informed that the monolayer structure of B-doped carbon in the form of BC_2 has more vacant sites above the Fermi energy state and thus provides a small barrier for the diffusion of Li and open circuit voltage among other forms of B-doped carbon [33]. Co-doping in graphene with several doping atoms, such as B/N doping, P/N doping, S/N doping, F/N doping, and Ge/N doping was further reported to enhance the reversible capacitance and performance efficiency. In commercial LIBs, mostly $LiCoO_2$ are used as cathodes and graphite as anodes. According to the theoretical calculation, the density of energy was obtained as ~387 Wh kg^{-1}, while the experimentally measured value of the energy density is observed from 120 Wh kg^{-1} to 150 Wh kg^{-1}. In the present time, the design of novel materials for electrodes for the purpose of large storage capacity, along with a high rate of capability in LIBs, is a big concern for researchers. For the design of the novel electrode materials, 2D nanosheets are implemented with nano-objects by engaging metal nanoparticles, atoms, and groups on the surface of the nanosheet. In the development of LIBs, few-layered metal dichalcogenides ($MoSe_2$, MoS_2, WSe_2, WS_2, ReS_2, TiS_2, and VS_2) are emerging as potential electrode materials [34]. Du *et al.* reported that MoS_2 nanosheet anodes in LIBs offer a higher storage capacity (theoretically ~1675 mAh g^{-1} and experimentally ~900 mAh g^{-1}) than graphite due to the larger coulomb efficiency of MoS_2 [35]. Whereas, a small value of electrical conductivity of MoS_2 leads to the low stability of the life cycle, as well as the quick degradation of the electrode material owing to the formation of polysulfide. The hybridization of MoS_2 nanosheets and RGO was reported as the solution to these problems. The RGO/MoS_2 hybrid electrode presented a capacity of ~2200 mAh g^{-1} along with a constant capacitance for more than 100 charging/discharging cycles. The capacitance value of TMD/graphene hybrids is regulated in the range of ~600–1546 mAh g^{-1}, subject to the synthesis process, composite, and quality of TMD materials. The WS_2/MoS_2 hybrid nanosheet was demonstrated as an anode material in LIBs to improve the motion of lithium ions and the binding of lithium ions on the electrode surface [36]. VS_2 is an example of metallic TMD, which has low barriers for diffusion. A single-layered VS_2 nanosheet was shown theoretically to steadily adsorb a few layers of Li, producing a capacity of ~1396 mAh g^{-1}. [37] Single-layer monochalcogenides (SiSe, GeSe, SnSe, etc.) were expected as anode in LIBs with low barrier heights of diffusion, modest to large specific capacitance, small open-circuit voltage, negligible change in volume, and high electrical conductivity. 2D transition metal oxides, like NiO, Co_3O_4, FeO_x, MnO_2, TiO_2, VO_x, and SnO_2, are extensively studied as anode materials in LIBs due to their high value of predicted/theoretical capacitance, low cost, environmental friendliness, and easy natural availability. Among various metal oxides, the Co_3O_4 nanostructure demonstrated outstanding stability and capability, while SnO_2, TiO_2, and VO_x 2D structures hold lower capacities as per theoretical calculations. Though SnO_2 and FeO_x produced high volume changes during charge/discharge, they suffered from severe capacitance loss due to the reduction of the anode. With first-principles calculation, Leong *et al.* reported that transition metal oxide XO_2 (X stands for Ni, Co, Mn) 2D sheets were feasible as cathodes in LIBs and SIBs [38]. They showed that Li and Na atoms efficiently transfer from stable sites to unstable sites on the plane surfaces of MnO_2 single layers to decrease the lattice expansion, offering the reduction of the voltage of batteries. As 2D binary metal oxide, Cr_2O_3, MoO_2, and ZnO were described as efficient materials for electrodes in rechargeable batteries. The porous structure of 2D nanosheets facilitates for efficient operation of the battery. Peng *et al.* reported highly porous $ZnMn_2O_4$ nanosheets, which showed a specific capacity of ~ 480.0 mAh g^{-1} over 1000 charge/discharge cycles at 800.0 mA g^{-1} along with high coulomb efficiency [39]. Although 2D metal oxide shows good performance as electrode materials in LIBs, the low electrical conductivity of these materials limits their extensive applications on a large scale, which may be overcome by the formation of a hybrid structure with graphitic materials. Pure MXenes, being good conductors, allow high mobility of lithium ions, leading to high energy storage capabilities, whereas the surface functionalization of MXenes with -OH or -F groups makes small in-band gaps in MXenes, which reduce the transportation of Li ions, leading to high value of irreversible capacitance and small reversible capacitance [40]. The functionalization of MXenes surface with metal ions and metal oxides improves the storage capacity of Li ions. 2D transition metal borides such as Fe_2B_2 and Mo_2B_2 have been reported to have small barriers for the diffusion of Li ions along with a high capacity for Li ion storage [41]. Owing to extraordinary electrical conductance, high hypothetical capacity, and low potentials near those of Li metal, 2D metal nitrides were observed as prospective electrode materials for batteries.

SIBs are a perfect alternative to LIBs. Similar to LIBs, 2D nanomaterials in the form of $NaXMO_2$ (X stands for Ni, Mn, or Co) were reported as proficient materials as cathodes in SIBs [38]. As the size of the Na ion is more than ~55 % higher than that of the Li ion, a higher size of the pore diameter (minimum space of ~0.37 nm) is essential for the transportation of Na ions on the surface of the electrode. For SIBs, graphitic materials cannot be used as anodes like LIBs due to their irreversible electrochemical reaction and low capacity. MXenes group materials have been observed as encouraging anode materials in SIBs with high energy storage capacity because of the extraordinary conductivity and rapid diffusion rates of the ions in the MXenes. The required potential for the attachment of Na^+ ion on MXenes surface is also possible to tune by changing the transition metal elements and the functional groups T_x (T_x: -OH, -H, -F) [42]. Few-layered Ti_2CT_x has been demonstrated as anode material in SIBs owing to a reversible capacitance of ~174 mAh g^{-1} with a voltage of 1.3 V [43]. Sr_2N and Ca_2N with their monolayer structure have been implemented for the prediction of the capacities with values of ~1137 and 282 mAh g^{-1}, respectively, with Na ion adsorption window of ~−0.23 and −0.13 eV and energy barriers of ion diffusion of 0.08 and 0.02 eV [44].

Though the development of 2D nanomaterials as electrode materials is growing interest in various types of battery technology, several serious issues need to be resolved for extensive applications of energy storage. A major issue for the LIBs/SIBs is the high value of initial irreversible capacity and bad coulomb efficacy owing to the creation of solid layers at the electrolyte-electrode interface. For 2D materials as electrodes, the degradation of electrolytes produces lithium carbonate and organic components, which cause safety issues. Moreover, 2D nanomaterials as electrodes lead to small packing density and energy densities of energy storage, which are not favorable for moveable/wearable electronic devices.

7.2.2 Materials Used in Energy Conversion

7.2.2.1 Oxygen Reduction Reaction (ORR)

ORR is an important process to generate electricity from hydrogen gas for renewable batteries as well as fuel cells. In ORR, oxygen (O_2) is deformed to H_2O in an acidic medium of electrolyte, whereas O_2 is broken into OH^- ions in a basic medium, and later, OH^- ions combine with an H^+ to produce water [7]. O_2 reduction comprises the generation of both O-O and O-H bonds. The ORR process can continue using $4e^-$ or $2e^-$ [7]. The reaction with $4e^-$ is required to get a high value of efficiency along with large open circuit voltage in fuel cells, whereas the $2e^-$ pathway is suitable for trade production of H_2O_2. The kinetics of ORR are fundamentally slow and thus require the transformation of multi-e^- and multi-H^+. [45] To reduce the overpotential and speed up the reaction process, efficient electrocatalysts are essential. For optimal catalysis activity, reactants, outcomes, and intermediates cannot be attached to the catalyst too weakly or too strongly. Cheng *et al.* reported Pt as the most efficient catalyst for ORR via the 4e– pathway [46]. Nevertheless, the high price along with insufficient availability of Pt obstruct its mass implementation. Hence, the discovery of low-cost and easily available metals or non-metal catalysts has become a big concern nowadays. Strain engineering suggests a substitute approach for the modification of the surface properties for efficient catalytic materials. The modification of the surface strain can extraordinarily upgrade the ORR activity in the TMD group. With $NbSe_2$ nanosheets, a minimum overpotential of ~0.43 V was reported in the ORR process for a tensile strain of ~5% [47]. Modifying the inherent strain of transition metal nanosheets by atomic-level control of their thickness allows for maximum catalytic reactivity. Density functional theory (DFT) calculations showed a reduction of the surface energy in the nanosheet, generating more compressive strains in the nanosheet plane. The upsurge of in-plane compressive strains weakens the adsorption energy of atomic oxygen, which enhances catalytic activity. Further, they experimentally confirmed the idea with Pd nanosheets for alkaline ORR. By the Linear Sweep Voltammetry (LSV) method, the half-wave potential was measured as 0.90 V, 0.95 V, and 0.93 V for the 8-layered,

5-layered, and tri-layered Pd nanosheets, respectively. The relation of the ORR activity with the thickness of Pd nanosheets was also elucidated by free energy barriers calculation for 3 to 8 molecular layers of Pd nanosheets. Rowley-Neale *et al.* demonstrated ORR activity in an acidic medium using MoS_2 with defect sites, and they observed the 4e$^-$ transfer process in the ORR activity, which was enhanced by the defect states in MoS_2 [48]. In MoS_2, the electro-positive Mo atoms at the edge sites interact with the electro-negative O atoms in the electrolyte, promoting ORR activity. The modification of the oxygen vacancies in 2D transition metal oxides endorses the adsorption of oxygen and consequently diminishes energy barriers by uncovering additional metal sites, which increases ORR catalysis. Though the intrinsic ORR process of TMDs with VI-B group metals is not significant, precise doping of MoS_2 by P or O induced improved adsorption of O_2 from the electrolyte, enhancing ORR with 4e$^-$ transfer. A DFT calculation showed that transition metal atoms could alter the electronic behaviors of monolayer MoS_2 by interacting with the vacant states of S atoms. The conductive behaviors of graphene can be drastically altered by heteroatom doping, which makes p- or n-type structures of graphene. Among various doping methods, N doping in graphene was investigated intensively for ORR catalysis. In N-doped graphene (NG), graphitic-N shows an n-type nature, whereas pyridinic, pyrrolic, and nitrilic N produce p-type material [49]. Both graphitic-N and pyridinic-N in NG were reported to encourage ORR activity. The pyridinic-N present in the plan of the surface owns a lone pair of electrons, which facilitates O_2 adsorption. On the other side, graphitic N transfers e$^-$ to the delocalized π state, resulting in the enhancement of the nucleophilic behavior of the neighboring carbon atoms and thus upholding ORR. Additionally, by the alteration of the electronic charge distribution in NG, the ORR process may be upgraded. The DFT computation and the experimental observation claimed that the ORR catalytic efficiency of doped graphene is highest for NG compared to other doping in graphene, and they follow the sequence NG > SG > PG > FG > IG > BG > BrG > ClG (G: Graphene) for onset potentials and current densities limit. Due to the low electrical conductivity of g-C_3N_4, it could not show significant ORR activity. However, the formation of a heterostructure with RGO has assisted g-C_3N_4 to behave as a superior catalyst for the ORR process, close to that of commercialized Pt/C catalysts in a basic electrolyte. For further improvement of ORR activities, the conductive characteristics and physical behaviors of 2D nanomaterials are required to be modified by the alteration of the type and position of the doped atom, co-doping with various heteroatoms, tailoring of strain, and generation of porous structures.

With a well-regulated synthesis and *in-situ* method, the efficiency of the electrolytes in acidic and alkaline media can be enhanced.

7.2.2.2 Oxygen Evolution Reaction (OER)

OER is one of the interesting processes in water electrolysis and metal-air batteries. OER activity follows a 4e- pathway in alkaline and acidic media using the following equations:

$$4OH^- = O_2 + 2H_2O + 4e^- \text{ in high pH medium,}$$

$$2H_2O = O_2 + 4H^+ + 4e^- \text{ in low pH/neutral medium}$$

OER usually involves a large value of overpotential for breaking O-H and forming O-O bonds. Similar to ORR, the optimization of the binding energy of the intimidate states in OER, such as OH*, O*, and OOH*, is crucial for high electrocatalytic activity. In OER, the performance of catalysts could be assessed by the required overpotential to reach a current density of ~10 mA cm^{-2}, which is about the current density for a solar cell with a conversion efficiency of 10%. The Tafel slope is another significant parameter that can give confirmation for the kinetics and mechanism of the electrocatalyst for the OER. For OER activity, the electrocatalysts RuO_2, IrO_2, and Ru-Ir oxides are commonly used in both high and low pH media, even though these costly rutile catalysts need overpotentials greater than 200 mV to get a 10 mA cm^{-2} current density [50]. In recent times, a number of 2D transition metals have been reported to be inexpensive and easily available, along with proficient catalytic activity, such as Fe-doped Ni_2P [51], oxyhydroxide (γ-CoOOH) [52], etc. Many of these reported non-noble metal electrocatalysts exceed the overpotentials value of 220 and 280 mV for chemically prepared RuO_2 and IrO_2, respectively, at 10 mA cm^{-2}. For the current density of 10 mA cm^{-2}, the reported overpotential of Fe-doped Ni_2P was ~190 mV [51], for γ-CoOOH was ~300 mV [52], and for amorphous CoFe hydroxide was ~280 mV [53]. In OER activity, MXenes nanosheets and graphene offer superior performance as catalysts that produce high electrical conduction and interesting surface properties. Chen *et al.* reported OER activity with a heterostructure of cobalt borate (CoB) and graphene, which offered an overpotential of ~290 mV along with a Tafel slope of ~53 mV dec^{-1} in a basic electrolyte at 10 mA cm^{-2} [54]. They also demonstrated a satisfactory performance of the graphene hybrid in a neutral medium with an onset overpotential of ~234 mV and a current density of ~14.3 mA cm^{-2} at 1.8 V. An assembly of Ti_3C_2 in MXene group and the porous form of g-C_3N_4 was presented as an efficient catalyst in the OER process with an onset overpotential ~210 mV [55]. The nitrogen-rich g-C_3N_4 proved itself as a stable and tunable electrocatalyst for efficient OER activity after the formation of heterostructure with MXenes to overcome its inherent low electrical conductivity. Nevertheless, metal-free electrodes have still not beat the performance of traditional electrodes for the OER process.

7.2.2.3 Hydrogen Evolution Reaction (HER)

The environment-friendly hydrogen (H_2) is a promising energy option with a high energy capacity of ~143 mJ kg^{-1} [56]. In recent dates, 2D nanomaterials have been getting huge attention as catalysts in hydrogen evolution reactions (HER) over metal-based catalysts. The HER reaction is given by the following equations: $2H^+ + 2e^- \rightarrow H_2$ in an acidic medium, and $2H_2O + 2e^- \rightarrow H_2 + 2OH^-$ in the basic electrolyte, which takes place in the cathode to produce H_2 from H_2O. In the acidic medium, H^+ is absorbed from the electrolyte onto the cathode surface and forms a hydrogen atom (H). In contrast, water offers H^+ for adsorption in alkaline media, being dissociated following the reaction $H_2O + e^- \rightarrow H^+ + OH^- + e^- \rightarrow H + OH^-$. Later, two H atoms combine to form an H_2 molecule [57]. A diverse collection of 2D transition metals has been developed as interesting catalysts in the HER process, which involves transition metals, metal chalcogenides, oxides, and hydroxides. Specifically, WO_3 nanosheets enriched with oxygen vacancies offered a tiny overpotential of ~38 mV at 10 mA cm^{-2} along with 38 mV dec^{-1} value of Tafel slope, which is a substitution of Pt in the case of HER electrocatalysis [58]. Metal chalcogenides usually show better conduction of charges, mechanical stability, thermal inertness, and superior electrochemical behavior than their associated oxide counterparts. Scientists are exploring exciting new materials of thin sheets of TMD groups, including MoS_2, $MoSe_2$, WSe_2, ReS_2, $MoTe_2$, etc. [7]. According to the previous study, the performance of selenide TMDs in HER was observed to be better than sulfides and tellurides, with $MoSe_2$ being the most proficient material [59]. The exact mechanism of how these materials work is still debated, but exposed edges seem to play a crucial role. Beyond TMDs, other materials are being investigated. Low-cost $NiSe_2$ nanosheets with high porosity are among the most promising 2D metal-catalysts working in acidic media [60]. In alkaline media, the core-shell hybrid of $Co(OH)_2$@polyaniline shows excellent performance, surpassing even Pt catalysts [61]. Graphene, while not ideal for HER on its own, can be improved by doping or co-doping with elements like B, N, or P. Combining doped graphene with materials like g-C_3N_4 can further enhance its activity [62]. Black phosphorus is another potential material, but its effectiveness is limited by strong hydrogen adsorption energy of ~1.01 eV [63]. Doping with metals can improve its performance by adjusting this binding energy. Although the 2D materials have been widely implemented for the advancements of HER, highly proficient electrocatalysis for HER in alkaline solution is still a big concern for researchers. Combining strategies like creating heterostructures, introducing pores, and manipulating electronic properties holds promise for further breakthroughs.

7.2.2.4 Carbon Dioxide Reduction Reaction (CRR)

Turning CO_2 into valuable products with clean electricity offers a promising solution to climate change. This process, called electrocatalytic CO_2 conversion, captures carbon emissions while storing renewable energy (like solar or wind power) as chemical energy. However, CO_2 is a tough nut to crack because it is very stable. Special catalysts are needed to get the conversion going. These catalysts work by breaking down the CO_2 molecule step-by-step, adding electrons and protons at each stage. Depending on how many electrons and protons are added, different products are formed, like carbon monoxide (CO), methanol (CH_3OH), or even ethanol (C_2H_5OH) [64]. There are four steps in this process: CO_2 sticking to the catalyst surface, activating the CO_2 to make it easier to break apart, breaking the carbon-oxygen bond and adding electrons and protons, and lastly, releasing the product from the catalyst. The first step is crucial, as CO_2 needs to be attracted to the catalyst for the conversion to happen. Scientists are developing new types of catalysts, especially those made from 2D materials like nanosheets, to improve this process for large-scale production. TMD nanosheets have shown excellent performance in converting CO_2 to CO, especially in ionic liquids like 1-Ethyl-3-methylimidazolium tetrafluoroborate (EMIM-BF_4). Researchers are also exploring alloys of these materials, which seem to be even more effective. While pristine graphene (pure sheets of carbon atoms) is not a good choice for CO_2 conversion, doping it with N and B can significantly improve its performance [65]. Similarly, metal-nitrogen doped carbon nanosheets show promise for CO_2 conversion to CO. Another interesting metal-free option is a material called g-C_3N_4. This material has special nitrogen atoms in its structure that attract CO_2 and produce oxygen-containing intermediate products like *OCHO, *OCH_x, *OH, and *O, which indicates high efficiency in CO_2 conversion [45]. Efficient electrochemical reduction of CO_2 remains a big challenge.

7.2.2.5 Water Splitting

Water splitting is used to generate H_2 energy by photocatalysis and photoelectrocatalytic process. Solar light is used to split water by the photocatalytic process with three crucial steps: first, an electron-hole (e-h) pair is generated by the excitation of a bandgap; then, charge carriers are separated and relocated to the photocatalyst surface; after that, electrons and holes generated by solar light are transferred to the reacting material to activate associated redox reaction [66]. The production of H_2 by the catalytic effect through the splitting of water molecules is crucial as environment-friendly H_2 gas has the maximum energy density among various carbon-based power sources. Due to their interesting intrinsic properties like high specific surface area, tunable band gap, and small migration distance of the charge carriers, 2D nanosheets have been explored as superior alternatives for photocatalysis. Nanosheets of metal oxides, such as TiO_2, MnO_2, WO_3, and $KTaO_3$, have been extensively explored as catalysts in water splitting because of their appropriate band alignment and stability during photocatalysis. 2D sulfide nanosheets like CdS, SnS, GaS, etc., have high valence band positions compared to those of S 3p states and the conduction band, facilitating the reduction of H_2O to H_2 [67]. Although sulfides have a better response to solar light than oxides, the photo-corrosive nature of sulfides limits their implementation. Jing *et al.* reported photocatalytic water splitting with ultrathin bismuth oxyhalide [68]. The formation of hybrid structures has also been reported for H_2 evolution by water splitting [69]. Although inordinate development has been made in water splitting with 2D photocatalysts, low production, high price, low stability, and slow charge transfer and trapping are the present issues in photocatalysis water splitting with 2D materials.

7.3 Summary

2D nanosheets have shown significant potential for energy conversion and electrodes. They have achieved benchmark-surpassing performance compared to traditional precious metal catalysts. Overall, nanosheet catalysis is becoming a powerful tool for sustainable energy conversion and storage owing to its effectiveness and reduced reliance on noble metals. Future discoveries of novel 2D materials and the development of high-performance electrodes with improved properties hold great promise for further advancements in this field. Creating new materials and structures of 2D scaffolds is essential to improve renewable energy sources. The formation of hybrids is a promising approach to overcome limitations and improve catalysts for complex processes. Tailoring 2D structures and their interactions is important for maximizing their effectiveness as catalysts. Large-scale, cost-effective production methods are needed for practical applications of these catalysts. Researchers need to improve methods for studying these catalysts and their mechanisms. Better understanding and control over surface chemistry are crucial for further performance gains.

References

1. Raza, S., *et al.*, Electrochemistry of 2D-materials for the remediation of environmental pollutants and alternative energy storage/conversion materials and devices, a comprehensive review. *Sustain. Mater. Technol.*, 40, e00963, 2024.
2. Qi, J., *et al.*, Emerging Two-Dimensional Materials for Proton-Based Energy Storage. *ACS Nano*, 18, 25910–25929, 2024.
3. Ren, Y. and Xu, Y., Recent advances in two-dimensional polymers: synthesis, assembly and energy-related applications. *Chem. Soc. Rev.*, 53, 1823–1869, 2024.
4. Mounet, N., *et al.*, Two-dimensional materials from high-throughput computational exfoliation of experimentally known compounds. *Nat. Nanotechnol.*, 13, 3, 246–252, 2018.
5. Bora, A., *et al.*, Understanding the excitation wavelength dependent spectral shift and large exciton binding energy of tungsten disulfide quantum dots and its interaction with single-walled carbon nanotubes. *J. Colloid Interface Sci.*, 561, 519–532, 2019.
6. Ramachandran, T., *et al.*, A review of 2D metal boride-derived nanostructures: From synthesis to energy storage and conversion applications. *J. Energy Storage*, 99, 113425, 2024.
7. Tao, H., *et al.*, Two-dimensional materials for energy conversion and storage. *Prog. Mater Sci.*, 111, 100637, 2020.
8. Das, R., *et al.*, Quantitative understanding of charge transfer mediated Fe3+ sensing and fast photoresponse by N-doped graphene quantum dots decorated on plasmonic Au nanoparticles. *ACS Appl. Mater. Interfaces*, 12, 4, 4755–4768, 2020.
9. Das, R., Paul, K.K., Giri, P.K., Highly sensitive and selective label-free detection of dopamine in human serum based on nitrogen-doped graphene quantum dots decorated on Au nanoparticles: mechanistic insights through microscopic and spectroscopic studies. *Appl. Surf. Sci.*, 490, 318–330, 2019.
10. Manzeli, S., *et al.*, 2D transition metal dichalcogenides. *Nat. Rev. Mater.*, 2, 8, 1–15, 2017.
11. Wang, H. and Qian, X., Two-dimensional multiferroics in monolayer group IV monochalcogenides. *2D Mater.*, 4, 1, 015042, 2017.

12. Pandey, M., *et al.*, Prospects of MXene and graphene for energy storage and conversion. *Renew. Sustain. Energy Rev.*, 189, 114030, 2024.
13. Bibi, F., *et al.*, Advances in 2D/2D MXenes-based heterostructures for energy storage/conversion applications. *J. Mater. Sci. Technol.*, 202, 82–118, 2024.
14. Wang, L., *et al.*, Tunable intrinsic strain in two-dimensional transition metal electrocatalysts. *Science*, 363, 6429, 870–874, 2019.
15. Zhu, J., *et al.*, Two-dimensional porous polymers: from sandwich-like structure to layered skeleton. *Acc. Chem. Res.*, 51, 12, 3191–3202, 2018.
16. Parveen, S., *et al.*, Large exciton binding energy, high photoluminescence quantum yield and improved photostability of organo-metal halide hybrid perovskite quantum dots grown on a mesoporous titanium dioxide template. *J. Colloid Interface Sci.*, 539, 619–633, 2019.
17. Yang, W., *et al.*, Graphene in supercapacitor applications. *Curr. Opin. Colloid Interface Sci.*, 20, 5-6, 416–428, 2015.
18. Xia, J., *et al.*, Measurement of the quantum capacitance of graphene. *Nat. Nanotechnol.*, 4, 8, 505–509, 2009.
19. El-Kady, M.F., *et al.*, Laser scribing of high-performance and flexible graphene-based electrochemical capacitors. *Science*, 335, 6074, 1326–1330, 2012.
20. Han, J., *et al.*, Generation of B-doped graphene nanoplatelets using a solution process and their supercapacitor applications. *ACS Nano*, 7, 1, 19–26, 2013.
21. Karim, G.M., *et al.*, Ultra-fast electro-reduction and activation of graphene for high energy density wearable supercapacitor asymmetrically designed with MXene. *Carbon*, 203, 191–201, 2023.
22. Levi, M.D., *et al.*, Solving the capacitive paradox of 2D MXene using electrochemical quartz-crystal admittance and *in situ* electronic conductance measurements. *Adv. Energy Mater.*, 5, 1, 1400815, 2015.
23. Zhang, C., *et al.*, Transparent, flexible, and conductive 2D titanium carbide (MXene) films with high volumetric capacitance. *Adv. Mater.*, 29, 36, 1702678, 2017.
24. Acerce, M., Voiry, D., Chhowalla, M., Metallic 1T phase MoS2 nanosheets as supercapacitor electrode materials. *Nat. Nanotechnol.*, 10, 4, 313–318, 2015.
25. Huang, J., *et al.*, Effect of Al-doped β-Ni(OH) 2 nanosheets on electrochemical behaviors for high performance supercapacitor application. *J. Power Sources*, 232, 370–375, 2013.
26. Huang, S., *et al.*, Immobilization of Co–Al layered double hydroxides on graphene oxide nanosheets: growth mechanism and supercapacitor studies. *ACS Appl. Mater. Interfaces*, 4, 4, 2242–2249, 2012.
27. Chen, H., *et al.*, One-step fabrication of ultrathin porous nickel hydroxide-manganese dioxide hybrid nanosheets for supercapacitor electrodes with excellent capacitive performance. *Adv. Energy Mater.*, 3, 12, 1636–1646, 2013.
28. Dudney, N.J. and Li, J., Using all energy in a battery. *Science*, 347, 6218, 131–132, 2015.

29. Yoon, D., *et al.*, Hydrogen-Enriched Reduced Graphene Oxide with Enhanced Electrochemical Performance in Lithium Ion Batteries. *Chem. Mater.*, 27, 1, 266–275, 2015.
30. Zhao, D., *et al.*, From graphite to porous graphene-like nanosheets for high rate lithium-ion batteries. *Nano Res.*, 8, 2998–3010, 2015.
31. Sahoo, M., *et al.*, Green synthesis of boron doped graphene and its application as high performance anode material in Li ion battery. *Mater. Res. Bull.*, 61, 383–390, 2015.
32. Hardikar, R.P., *et al.*, Boron doped defective graphene as a potential anode material for Li-ion batteries. *PCCP*, 16, 31, 16502–16508, 2014.
33. Das, D., *et al.*, Monolayer BC 2: an ultrahigh capacity anode material for Li ion batteries. *PCCP*, 19, 35, 24230–24239, 2017.
34. Yu, X.Y. and Lou, X.W., Mixed metal sulfides for electrochemical energy storage and conversion. *Adv. Energy Mater.*, 8, 3, 1701592, 2018.
35. Du, G., *et al.*, Superior stability and high capacity of restacked molybdenum disulfide as anode material for lithium ion batteries. *Chem. Commun.*, 46, 7, 1106–1108, 2010.
36. Wang, D., *et al.*, Potential application of metal dichalcogenides double-layered heterostructures as anode materials for Li-ion batteries. *J. Phys. Chem. C*, 120, 9, 4779–4788, 2016.
37. Wang, D., *et al.*, Two-dimensional VS 2 monolayers as potential anode materials for lithium-ion batteries and beyond: first-principles calculations. *J. Mater. Chem. A*, 5, 40, 21370–21377, 2017.
38. Leong, C.C., Pan, H., Ho, S.K., Two-dimensional transition-metal oxide monolayers as cathode materials for Li and Na ion batteries. *PCCP*, 18, 10, 7527–7534, 2016.
39. Peng, L., *et al.*, Holey two-dimensional transition metal oxide nanosheets for efficient energy storage. *Nat. Commun.*, 8, 1, 15139, 2017.
40. Xie, Y., *et al.*, Role of Surface Structure on Li-Ion Energy Storage Capacity of Two-Dimensional Transition-Metal Carbides. *J. Am. Chem. Soc.*, 136, 17, 6385–6394, 2014.
41. Guo, Z., Zhou, J., Sun, Z., New two-dimensional transition metal borides for Li ion batteries and electrocatalysis. *J. Mater. Chem. A*, 5, 45, 23530–23535, 2017.
42. Kajiyama, S., *et al.*, Sodium-ion intercalation mechanism in MXene nanosheets. *ACS Nano*, 10, 3, 3334–3341, 2016.
43. Wang, X., *et al.*, Pseudocapacitance of MXene nanosheets for high-power sodium-ion hybrid capacitors. *Nat. Commun.*, 6, 1, 6544, 2015.
44. Hu, J., *et al.*, 2D electrides as promising anode materials for Na-ion batteries from first-principles study. *ACS Appl. Mater. Interfaces*, 7, 43, 24016–24022, 2015.
45. Jin, H., *et al.*, Emerging two-dimensional nanomaterials for electrocatalysis. *Chem. Rev.*, 118, 13, 6337–6408, 2018.

46. Cheng, H., *et al.*, Catalytic system based on sub-2 nm Pt particles and its extraordinary activity and durability for oxygen reduction. *Nano Lett.*, 19, 8, 4997–5002, 2019.
47. Zhao, S., *et al.*, Group VB transition metal dichalcogenides for oxygen reduction reaction and strain-enhanced activity governed by p-orbital electrons of chalcogen. *Nano Res.*, 12, 925–930, 2019.
48. Rowley-Neale, S.J., *et al.*, 2D molybdenum disulphide (2D-MoS 2) modified electrodes explored towards the oxygen reduction reaction. *Nanoscale*, 8, 31, 14767–14777, 2016.
49. Das, R., *et al.*, Origin of high photoluminescence yield and high SERS sensitivity of nitrogen-doped graphene quantum dots. *Carbon*, 160, 273–286, 2020.
50. Audichon, T., *et al.*, IrO_2 coated on RuO_2 as efficient and stable electroactive nanocatalysts for electrochemical water splitting. *J. Phys. Chem. C*, 120, 5, 2562–2573, 2016.
51. Wang, J., *et al.*, Fe-doped Ni2P nanosheet array for high-efficiency electrochemical water oxidation. *Inorg. Chem.*, 56, 3, 1041–1044, 2017.
52. Huang, J., *et al.*, CoOOH nanosheets with high mass activity for water oxidation. *Angew. Chem.*, 127, 30, 8846–8851, 2015.
53. Han, X., *et al.*, Mass and charge transfer coenhanced oxygen evolution behaviors in CoFe-layered double hydroxide assembled on graphene. *Adv. Mater. Interfaces*, 3, 7, 1500782, 2016.
54. Chen, P., *et al.*, Strong-coupled cobalt borate nanosheets/graphene hybrid as electrocatalyst for water oxidation under both alkaline and neutral conditions. *Angew. Chem. Int. Ed.*, 55, 7, 2488–2492, 2016.
55. Ma, T.Y., *et al.*, Interacting carbon nitride and titanium carbide nanosheets for high-performance oxygen evolution. *Angew. Chem. Int. Ed.*, 55, 3, 1138–1142, 2016.
56. Subbaraman, R., *et al.*, Enhancing hydrogen evolution activity in water splitting by tailoring Li+-Ni (OH) 2-Pt interfaces. *Science*, 334, 6060, 1256–1260, 2011.
57. Conway, B. and Tilak, B., Interfacial processes involving electrocatalytic evolution and oxidation of H2, and the role of chemisorbed H. *Electrochim. Acta*, 47, 22-23, 3571–3594, 2002.
58. Zheng, T., *et al.*, Conductive tungsten oxide nanosheets for highly efficient hydrogen evolution. *Nano Lett.*, 17, 12, 7968–7973, 2017.
59. Gholamvand, Z., *et al.*, Comparison of liquid exfoliated transition metal dichalcogenides reveals MoSe 2 to be the most effective hydrogen evolution catalyst. *Nanoscale*, 8, 10, 5737–5749, 2016.
60. Liang, H., *et al.*, Porous Two-Dimensional Nanosheets Converted from Layered Double Hydroxides and Their Applications in Electrocatalytic Water Splitting. *Chem. Mater.*, 27, 16, 5702–5711, 2015.

61. Feng, J.X., *et al.*, $Co(OH)_2$@ PANI hybrid nanosheets with 3D networks as high-performance electrocatalysts for hydrogen evolution reaction. *Adv. Mater.*, 27, 44, 7051–7057, 2015.
62. Duan, J., *et al.*, Porous C3N4 nanolayers@ N-graphene films as catalyst electrodes for highly efficient hydrogen evolution. *ACS Nano*, 9, 1, 931–940, 2015.
63. Wei, Y. and Sun, Z., Liquid-phase exfoliation of graphite for mass production of pristine few-layer graphene. *Curr. Opin. Colloid Interface Sci.*, 20, 5, 311–321, 2015.
64. Sun, Z., *et al.*, Fundamentals and Challenges of Electrochemical CO2 Reduction Using Two-Dimensional Materials. *Chem*, 3, 4, 560–587, 2017.
65. Liu, Y., Zhao, J., Cai, Q., Pyrrolic-nitrogen doped graphene: a metal-free electrocatalyst with high efficiency and selectivity for the reduction of carbon dioxide to formic acid: a computational study. *PCCP*, 18, 7, 5491–5498, 2016.
66. Ghosh, R., *et al.*, Multifunctional Ag nanoparticle decorated Si nanowires for sensing, photocatalysis and light emission applications. *J. Colloid Interface Sci.*, 532, 464–473, 2018.
67. Sun, Y., *et al.*, All-surface-atomic-metal chalcogenide sheets for high-efficiency visible-light photoelectrochemical water splitting. *Adv. Energy Mater.*, 4, 1, 1300611, 2014.
68. Jing, T., *et al.*, The photocatalytic properties of ultrathin bismuth oxychloride nanosheets: a first principles study. *PCCP*, 18, 10, 7261–7268, 2016.
69. Meng, F., *et al.*, Solar Hydrogen Generation by Nanoscale p–n Junction of p-type Molybdenum Disulfide/n-type Nitrogen-Doped Reduced Graphene Oxide. *J. Am. Chem. Soc.*, 135, 28, 10286–10289, 2013.

8

Rare Earth Elements in Photonic Materials

M. Dhamodhara Naidu[1]* and Kolla Bhanu Prakash[2]

[1]*Department of Physics, St. Martin's Engineering College, Secunderabad, Telangana, India*
[2]*Department of Computer Science & Engineering, Koneru Lakshmaiah Education Foundation, Vaddeswaram, India*

Abstract
This chapter provides an overview of rare earth (RE) elements, their history, electronic configuration, and different properties like physical, chemical, and optical properties. Various rare earth-doped materials and their applications are explained in the field of luminescent devices. The RE elements in the periodic table are a set of 17 metallic elements. These comprise the chemical elements scandium, with an atomic number of 21, yttrium, with an atomic number of 39, and 15 lanthanides, with atomic numbers from 57 to 71 (Lanthanum to Lutetium). The properties of rare earth elements, like physical and chemical properties, are influenced by their electronic configuration. The general electronic configuration is $[Xe]4f^n$, where n varies from 1 to 14, corresponding to different RE elements. In general, RE elements are in a trivalent state and this state of RE ions is the most stable ionization state. Rare earth ions have chemical similarity so they can be easily substituted in different oxide compounds such as B_2O_3, P_2O_5, SiO_2, and GeO_2, known as rare earth-doped borate, phosphate, silicate, and germinate glasses, respectively. Rare earth elements have shown interesting optical properties when incorporated into different oxide glass materials and phosphors. Depending on their use in leading technologies, RE-doped materials are used in applications such as field emission displays, solid-state lighting, television screens, plasma display panels, electroluminescent devices, and laser emitting devices.

Keywords: Rare earth elements, optical properties, photonics, luminescent devices

**Corresponding author*: dhamu.muppiri@gmail.com

Kolla Bhanu Prakash, S. V. Ranganayakulu and K. S. Jagannatha Rao (eds.) New Frontiers in Materials Science: Interdisciplinary Approaches to Innovation and Technologies, (141–158) © 2025 Scrivener Publishing LLC

8.1 Introduction

Rare earth (RE) elements constitute a distinct group of 17 elements in the periodic table, particularly the lanthanides and two more elements, scandium and yttrium [1]. The lanthanides series consists of a group of 15 elements, but along with scandium and yttrium, these are known as RE elements because they are extracted from the same ore deposits as the lanthanides. Almost all RE elements have the same chemical properties, but generally, lanthanides exhibit distinct luminescent characteristics and are regarded as non-toxic elements [2, 3]. Their emission profiles cover wide spectral ranges of wavelengths such as ultraviolet (UV) (100 to 400 nm), visible (400 to 700 nm), and near infrared (NIR) (700 to 1400 nm) regions. So, RE elements are perfect candidates for applications like laser-emitting materials, solid-state lighting, and optical fibers. Lanthanide elements can be classified into two types: (i) the elements lanthanum, yttrium, scandium, and lutetium, which have closed shells and their first excited energy levels exist in the far ultraviolet; and (ii) The other elements from praseodymium to ytterbium, containing 2 to 13 electrons, which have their first excited energy levels presented in the infrared region.

Because of the unusual spectroscopic properties of RE elements and due to the intraconfigurational f-f transitions, they were used to examine the variations of local structures in the host glass materials [4]. Among different RE-doped materials, the excited state lifetimes of the RE ions are almost up to 10 ms, and they show the property of exceptional fluorescence efficiency. This property makes lanthanides ideal active ions in most of the glass lasers. Because of the electronic configuration and the unfilled $4f^n$ orbitals, different glass lasers use lanthanide elements as active ions in the RE-doped glasses having suitable energy level schemes and selection rules for laser action. Different laser transitions are possible by the $4f^n$ energy level series of rare earth ions in the glass materials. The RE ions doped glass materials have many important applications some of them are fluorescent display devices, optical detectors, optical waveguides, optical amplifiers, and laser-emitting materials [5–7]. Compared to different RE-doped materials, glasses doped with RE elements have a lot of advantages due to the ease of preparation of materials in terms of big rods, shaping the material, minimizing optical inhomogeneity, uniform optical properties, enhanced laser performance, and isotropic characteristics. Also, when compared to crystals and other RE-doped materials, the glasses doped by Ln^{3+} ions incorporated with a large amount.

Generally, the optical and luminescent properties of RE ions doped materials are influenced by the selected host materials. To get suitable industrial applications, the host glass materials play an important. They must have good thermal stability with a low melting temperature when prepared with heavy metals, high refractive index, and good chemical stability. The host material properties depend on the concentration of Ln^{3+} ions. A small change in the concentration of Ln^{3+} ions in some materials makes large changes in their properties, and also the properties depend on the type of dopant in the material. Glass materials doped with lanthanide elements are promising materials due to their more general optical properties [8–15]. As expected, the crystal spectra of Ln^{3+} ions doped materials, which have remarkable sharpness of optical transitions disappear. Generally, glasses doped with Ln^{3+} ions produce optical spectra that are inhomogeneously broadened. The lasers in glass materials have some tunability, which is less than or equal to 10%, and this is impossible in hosts like crystalline materials. In crystalline hosts, not much difference has been identified between transition rates and lifetimes. Among different host materials for RE ions, oxide glasses are considered perfect host materials because they have some natural advantages like thermal stability, chemical durability, and optical transparency for excitation and laser emission wavelengths [16].

When Ln^{3+} ions are incorporated in glasses and inorganic materials with low symmetry, the behavior of Ln^{3+} ions in both materials is similar but differs in the case of optical spectra for inhomogeneous broadening in glasses, which is due to the multiplicity of Ln^{3+} sites [17, 18]. The absorption and fluorescence phenomena have been observed in Ln^{3+} ions doped glass materials, and forced electric-dipole transition occurs due to lower symmetry structure. For the excitation process, optical pumping has been used with a xenon flash lamp as a broadband source, and due to the inhomogeneous nature of Ln^{3+} ions in glass material, most of the light can be absorbed. Moreover, the excited state population becomes less due to enhanced spontaneous emission probability driven by effective cross-section. Because of the above properties, glass materials have greater storage capacity compared to crystalline materials.

8.2 History of Rare Earth Elements

The name RE elements is mainly based on their findings. The majority of RE elements were found during the 19th century, except yttrium, lutetium, and promethium, which were found in the years 1794, 1907, and 1943, respectively. Yttrium was found by chemist Johan Gadolin and a

Table 8.1 The origin of the names of rare earth ions.

Lanthanide	Discoverer	Year	Origin of its name
Yttrium	Gadolin	1794	Named after the town of Ytterby, Sweden.
Lanthanum	Mosander	1839	Based on the Greek word "lanthanon" meaning "I am hidden."
Cerium	Berzelius, Hisinger	1804	After the Greek deity of fertility, Ceres.
Praseodymium	Von Welsbach	1885	Based on the Greek words prasios (green) and didymos (twin).
Neodymium	Von Welsbach	1885	Based on the Greek word "neo," which means "new one."
Promethium	Marinsky, Coryell	1948	From Prometheus, who brought fire to mortals.
Samarium	De Boisbaudrau	1879	Named after Vasili Samarsky-Bykhovets, who found the RE element's ore called samarskite.
Europium	Demarcay	1901	Named after the continent Europe.
Gadolinium	De Boisbaudrau	1886	Based on Johan Gadolin (1760-1852) to honor his investigation of RE elements.
Terbium	Mosander	1843	Named after the village of Ytterby, Sweden.
Dysprosium	De Boisbaudrau	1886	Based on the Greek word "dysprositos," which means "hard to get."
Holmium	J.L. Soret	1878	Form the Latin word "Holmia" (Stockholm).
Erbium	Mosander	1843	Named after the town of Ytterby, Sweden.
Thulium	Cleve	1879	From the mythological land of Thule.
Ytterbium	De Marignac	1878	Named after the town of Ytterby, Sweden, where the first RE deposit was discovered.
Lutetium	Georges Urbain	1907	From Lutetia (ancient name of Paris).

Finnish mineralogist in a mineral later named gadolinite in his honor. The element gadolinium was named based on the mineral name gadolinite, which is indirectly named after the scientist Gadolin [19]. Another naturally occurring element, samarium, was named based on the mineral name samarskite. This element was named by Russian mining engineer, Vasili Samarsky-Bykhovets, the founder of the mineral. One RE element deposit was found in the 19th century in a deep pit or a place near the town of Ytterby in Sweden. These are known to be rare, and they are earth elements.

Table 8.2 Properties of RE elements.

Element	Atomic number	Atomic weight in a.m.u	Density in gcm^{-3}	Melting point in °C	Vicker's hardness (10kg load, kg/ mm^2)
Lanthanide	21	44.95	2.989	1541	85
Yttrium	39	88.90	4.469	1522	38
Lanthanum	57	138.90	6.146	918	37
Cerium	58	140.11	8.160	798	24
Praseodymium	59	140.90	6.773	931	37
Neodymium	60	144.24	7.008	1021	35
Promethium	61	145.00	7.264	1042	-
Samarium	62	150.36	7.520	1074	45
Europium	63	151.96	5.244	822	17
Gadolinium	64	157.25	7.901	1313	57
Terbium	65	158.92	8.230	1356	46
Dysprosium	66	162.50	8.551	1412	42
Holmium	67	164.93	8.795	1474	42
Erbium	68	167.26	9.066	1529	44
Thulium	69	168.93	9.321	1545	48
Ytterbium	70	173.04	6.966	819	21
Lutetium	71	174.97	9.841	1663	77

Initially, these RE elements were extracted as oxides. In French, an element in oxides was called "terre," which means "earth." In another scientific word in German at that time, the oxide form of this element was designated as the "erde," simply meaning "earth" [20, 21]. Table 8.1 presents information about the discovery of RE elements and their origins for names.

All RE elements were finally identified in the time of 20th century, but until 1945, the element promethium, which is the rarest, was not found, and until 1953, the metal lutetium was not purified [22]. Lanthanides are commonly classified into two groups: the first group is the light RE elements, which are lanthanum (57) to europium (63), and the second group is the heavy RE elements, which are gadolinium (64) to lutetium (71). Some selected properties of RE elements are noted in Table 8.2 [23]. In general, all rare earth elements are trivalent except cerium, which is available in Ce^{+4} forms. The elements europium, ytterbium, and samarium can form +2 ionic states in some environments, but these two have the same ionic radii. The chemical and physical differences among the RE elements are due to small differences in ionic radius. The similar radii and oxidation states of rare earth elements allow them for easy substitution into various crystal lattices. Among all RE elements, one element has no stable isotopes; very long half-life and limited stability are considered to be stable isotopes. The element promethium has no stable isotopes; it is radioactive and is not available as a free element on Earth. It is not a genuine element and is found in very minute quantities in natural materials.

8.3 Trivalent Lanthanides Electronic Configuration

Lanthanides have a common feature of electronic configuration based on the Xe atomic structure ($1s^2 2s^2 2p^6 3s^2 3p^6 4s^2 3d^{10} 4p^6 5s^2 4d^{10} 5p^6$) with the outer electrons ($6s^2$ or $6s^2 5d^1$) and n number of electrons in the 4f orbital, where $n \leq 14$. The electrons $6s^2 5d^1$ are removed when lanthanide ions are ionized and $[Xe]4f^n$ is the electronic configuration of triply ionized lanthanide elements. The trivalent state of lanthanides is the most stable ionization state, with electrons in the 5s and 5p orbitals, but the electrons in the outer orbitals shield the 4f energy states, reducing their interaction with external forces surrounding them. Table 8.3 shows the lanthanides and their electronic configuration along with ground and valence energy levels.

Table 8.3 The neutral lanthanides' electronic configurations with their ground and valence levels.

Lanthanides and symbol	Atomic number	Neutral atom electronic configuration	Valence states	Trivalent atom electronic configuration [Xe]$4f^n$, where n is	Ground state
Lanthanum (La)	57	[Xe] $4f^0\,5d^1\,6s^2$	3	0	$^1S_0$
Cerium (Ce)	58	[Xe] $4f^2\,6s^2$	3, 4	1	$^2F_{5/2}$
Praseodymium (Pr)	59	[Xe] $4f^3 6s^2$	3	2	$^3H_4$
Neodymium (Nd)	60	[Xe] $4f^4\,6s^2$	3	3	$^4I_{9/2}$
Promethium (Pm)	61	[Xe] $4f^5\,6s^2$	3	4	$^5I_4$
Samarium (Sm)	62	[Xe] $4f^6\,6s^2$	2, 3	5	$^6H_{5/2}$
Europium (Eu)	63	[Xe] $4f^7\,6s^2$	2, 3	6	$^7F_0$
Gadolinium (Gd)	64	[Xe] $4f^7\,5d^1\,6s^2$	3	7	$^8S_{7/2}$
Terbium (Tb)	65	[Xe] $4f^9\,6s^2$	3, 4	8	$^7F_6$
Dysprosium (Dy)	66	[Xe] $4f^{10} 6s^2$	3	9	$^6H_{15/2}$
Holmium (Ho)	67	[Xe] $4f^{11}\,6s^2$	3	10	$^5I_8$
Erbium (Er)	68	[Xe] $4f^{12}\,6s^2$	3	11	$^4I_{15/2}$
Thulium (Tm)	69	[Xe] $4f^{13}\,6s^2$	3	12	$^3H_6$
Ytterbium (Yb)	70	[Xe] $4f^{14}\,6s^2$	2, 3	13	$^2F_{7/2}$
Lutetium (Lu)	71	[Xe] $4f^{14}\,5d^1\,6s^2$	3	14	$^1S_0$

8.4 Energy Level Splitting of 4f States in Lanthanides

In lanthanide-doped materials, the crystal field, Coulombic, and spin-orbit interactions collectively remove the degeneracy of lanthanide ions energy levels. As shown in Figure 8.1, the energy level splitting of the $4f^n$

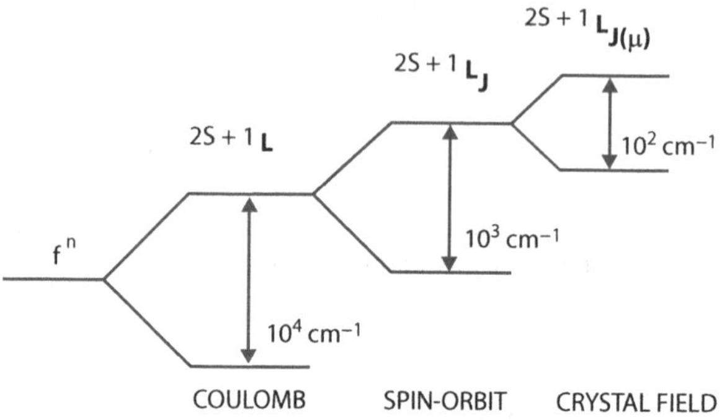

Figure 8.1 The 4f free-ion energy level degeneracy in the lanthanide.

configuration yields ^{2S+1}L multiplets with (2L+1)(2S+1) fold degeneracy, forming the basis of free ion energy levels.

The electrostatic field splits $^{2S+1}{}_{L}$ terms into energy levels separated by approximately 10^4 cm^{-1}. The spin-orbit interaction arises due to the interaction between the electron's intrinsic spin and the magnetic field induced by its orbital motion around the nucleus. Due to spin-orbit coupling, the $^{2S+1}{}_{L}$ terms are subdivided into (2J+1) energy levels designated as $^{(2S+1)}L_{J}$, This splitting typically yields energy differences of 10^3 cm^{-1}. For even-numbered electron systems (n), the interaction of 4f electrons interaction with the crystal field results in Stark splitting for $^{(2S+1)}L_J$ terms into (2J+1) distinct energy levels or the splitting of $^{(2S+1)}L_{J,\mu}$ energy terms into (J+1/2) energy levels if the electron number 'n' is odd. Normally, the Stark splitting observed in lanthanides, with energy differences of about 10^2 cm^{-1}, arises from the interplay of crystal field and spin-orbit coupling, which is approximately 10^3 cm^{-1}, greater than the crystal field interaction, which is approximately 10^2 cm^{-1}. Due to the shielding of 4f electrons by outer electrons, lanthanides show properties unlike those of transition metal ions.

8.5 The Energy Levels of Lanthanides and Dieke Diagram

The electronic structure of lanthanides is largely unaffected by the surrounding ion environment due to the shielding of the 4f electrons by the outer 5s and 5p electrons, but these energy levels, like 4f states, are simply the characteristics of Ln^{3+} ions. When RE ion is doped in host materials, the energy levels and 4f-4f transition energies of a particular ion are relatively unchanged. Figure 8.2 shows the Dieke diagram. Using this diagram, the characteristic energy state ${}^{2S+1}L_J$ of lanthanides doped into $LaCl_3$ was measured and this was reported by Dieke *et al.* [24, 25] in 1960, which remains a fundamental reference. For different host materials doped with trivalent lanthanides, the results of the energy levels of ${}^{2S+1}L_J$ multiplets in the Dieke diagram are used as reference work for any Ln^{3+} ions doped materials. The extension of the Dieke diagram by Ogasawara *et al.* [26] to higher energies revealed that the multiplet centers coincide with the energy levels of free ions in Russell-Saunders ${}^{2S+1}L_J$ coupling schemes, and the broadening of each state represents the splitting of energy levels due to the interaction of crystal fields with the ion's electronic configuration.

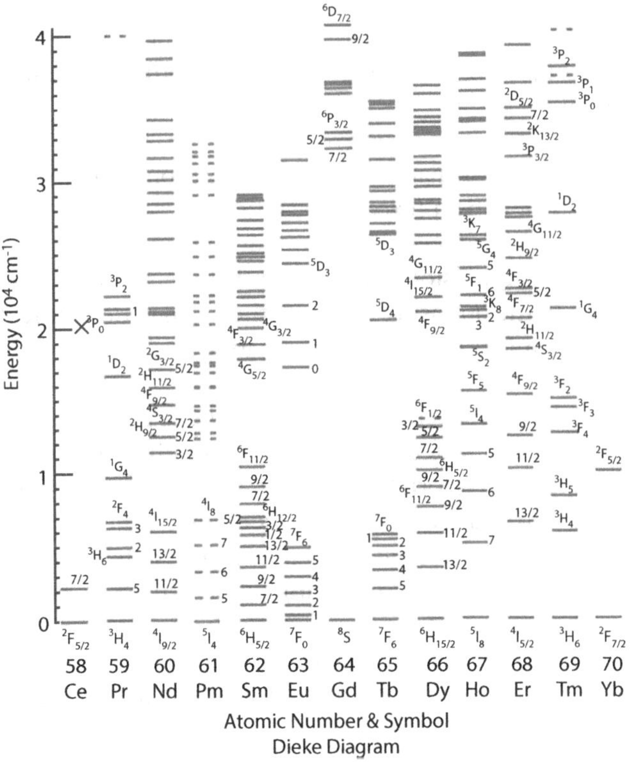

Figure 8.2 Energy levels of free RE^{3+} ions (Dieke diagram).

8.6 Physical and Chemical Properties of RE Elements

8.6.1 Basic Properties of Lanthanides

i. Lanthanides are grey or silvery white metals, when exposed to air, they are formed into their oxides.
ii. Lanthanides are generally soft metals. At higher atomic numbers, the hardness becomes slightly increased.
iii. Lanthanides have high melting and boiling points.
iv. Lanthanides have very little difference in solubility and complex formation.
v. When they react with water hydrogen is released (H_2), this reaction depends on temperature. The temperature influences lanthanide reactivity, with slower binding at lower temperatures and faster binding upon heating.
vi. At high temperatures in the air, lanthanides can burn easily.
vii. Lanthanides are strong reductants.
viii. Generally, lanthanide compounds are ionic and these show paramagnetism (except La^{3+} and Lu^{3+}).
ix. They have high electrical conductivity.
x. There is repulsion between Lanthanide ions and iron ions due to the opposing nature of their magnetic moments.
xi. Lanthanides have high coordination numbers, usually 8 or 9.
xii. Lanthanide spectroscopic and magnetic properties remain largely unaffected by the ligand coordination due to the shielding of 4f orbitals by outer $5s^2$ and $5p^6$ electrons.
xiii. The introduction of lanthanides into non-metallic elements, such as oxygen and nitrogen, results in brittleness, whereas calcium co-contamination enhances the corrosion propensity.
xiv. They do not form metallic bonds like Ln=O (or) Ln=N, which are known for many transition metals and some actinides.

Most of the RE elements have 3+ oxidation states, which is generally the most stable state, and all the rare earth oxides occur in the form RE_2O_3. Some lanthanides have different valence states. For example, praseodymium oxide, with the formula Pr_6O_{11}, usually contains 3+ and 4+ states depending on the conditions of formation. Similarly, Tb_4O_7 contains Tb^{4+}

and the more stable state Tb^{3+}, and Ce_2O_3 has Ce^{3+}, which is the most stable oxidation state. For all lanthanides, the valence states and their ground states are shown in Table 8.3.

8.6.2 Lanthanide Contraction

In the lanthanide series, it is noticed that the atomic radii and their different states of ion radii decrease when moving from lower atomic numbers to higher. This phenomenon is called lanthanide contraction. In this way, the atomic radius of lanthanum is the maximum, whereas lutetium has the minimum atomic radius within the lanthanide series. Lanthanide contraction happens because the inner 4f electrons don't effectively shield the outer electrons from the nuclear charge, causing the atoms to shrink in size. The valence electrons are not fully shielded from the nuclear attraction by inner electrons, resulting in varying degrees of shielding across periods and moving from left to right in the period of the periodic table. The imperfect shielding of valence electrons leads to an increase in the effective nuclear charge experienced by the outermost electrons. This effect is very noticeable, and the trend repeats, resulting in periodic patterns. There are 14 elements in this series, and the size of the lanthanides is determined by their 5s and 5p orbitals, as the 4f orbitals are too compact to significantly contribute to the atomic radius [27]. In general, the binding energy of electrons to the nucleus increases with atomic mass, leading to a stronger nuclear attraction, which in turn reduces ionic radius due to the enhanced electrostatic pull [27].

In the RE elements, the ionic radius of yttrium is comparable to heavier rare earth elements because yttrium's outermost electronic configuration resembles that of the heavy lanthanides, featuring a similar arrangement of electrons in the outermost energy levels. So, yttrium behaves chemically the same as the heavy RE elements [28]. In the other way, the atomic and ionic radius of scandium is substantially smaller than those of heavy RE earth elements, owing to its smaller atomic number and weaker shielding effect. Unlike other rare earth elements, scandium is not typically found in rare earth minerals, requiring separate extraction processes, and its chemical behavior is completely different from the other RE elements [23].

8.6.3 Chemical Properties

Rare earth elements are electropositive and form ionic compounds. Mineralogically, in the same way, they form halides, oxides, carbonates, silicates, phosphates, and arsenates or borates, but they do not form sulfides [29].

The different chemical properties of rare earth elements with air, nitrogen, hydrogen, carbon, silicon, and water are explained as follows.

Air: At elevated temperatures and when air is moist, the rare earth elements undergo oxidation rapidly. When the relative humidity increases from 1 to 75 %, the oxidation process increases by ten times [23].
Nitrogen: Though the reaction of RE elements with nitrogen is slow, at high temperatures the reaction rate is visible. The stable layer formed on the surfaces of RE elements prevents further nitridation [23].
Hydrogen: At temperatures from 400 to 600°C, rare earth metals can form hydrides, but many of them decompose easily [23].
Carbon: Rare earth elements easily form dicarbides (RE_2C_2). The lanthanides from lanthanum to samarium and gadolinium to holmium also form sesquicarbides (RE_2C_3). With nitrogen and oxygen, RE carbides form solid solutions [23].
Silicon: RE elements form silicides and other solid solutions, generally these are disilicides, which are $RESi_2$ [23].
Water: The reactions of RE elements with water can vary differently for different metals. At room temperature, the lighter rare earth elements reaction is slow, and at high temperature, the reaction becomes stronger. The heavy RE elements react slowly with water [23].

8.6.4 Magnetism

The Curie temperatures of lanthanide elements are usually low. At room temperature, these metals show paramagnetism except gadolinium, which has a Curie temperature of 292 K, which is 18.9°C, but it is a ferromagnetic material. From the reported data [30], terbium has a high Curie temperature of 222 K (-51.5°C). Next, some elements have their Curie temperatures as 89 K, 20 K, 20 K, and 32 K for dysprosium, holmium, erbium, and thulium, respectively.

8.7 Optical Properties

8.7.1 Luminescence

The emission of light by a body when it is heated is called incandescence. For example, when iron is heated, it emits red-yellow colored light. However, the term luminescence is not completely related to heat; it is electromagnetic radiation emitted by a body. The phenomenon of luminescence was identified and defined by English physicist G. C. Stokes and is called Stokes' law. As per Stokes' law, the wavelength of fluorescence or emission is shifted to a longer value compared to the exciting radiation. The phenomenon of luminescence was explained by German physicist E. Wiedemann. According to him, the light emission of certain kinds of materials absorbs various energies by the atoms in it and the luminescence can be explained based on the excitation process. Some of the luminescence phenomena like electroluminescence by electrons, photoluminescence by photons, cathodoluminescence by electron-induced current, thermo-luminescence by heat, etc., are possible [31].

The luminescence of lanthanides ranges the spectral region from UV to IR. All the lanthanides have unique optical and spectral properties in the visible region of the electromagnetic spectrum and they have different luminescent colors [32]. The applications of the lanthanides depend on their optical properties, especially these are based on the emission characteristics for color reproduction and lighting. Phosphors and LEDs are the most popular applications in photonics. Lanthanide ions convert ultraviolet (UV) or infrared (IR) excitation energies into visible radiation, detected by photosensitive instruments, such as solar cells [32]. For luminescence properties, europium and terbium are the most used lanthanides, and they emit red and green colors, respectively. Dysprosium exhibits an interesting luminescence with yellow color light, and samarium emits orange color light, but those emissions become weaker compared to europium and terbium. Among the lanthanides, the ions ytterbium, erbium, and neodymium emit radiation in the NIR region. The emission spectrum of holmium ions displays two distinct peaks: one in the red visible range and another in the infrared range. Europium ions in the divalent state exhibit blue emission, contributing to the visible spectrum [32].

The process of absorption of energy, fluorescence, and phosphorescence phenomenon can be explained as follows. Luminescence phenomena are broadly classified into fluorescence and phosphorescence, distinguished by their distinct emission lifetimes. Fluorescence is characterized by rapid light emission with a decay time of less than 10 milliseconds, while phosphorescence stands for the emission of light having a longer decay time, with values greater than 0.1s [31, 33]. Fluorescence and phosphorescence involve energy relaxation, resulting in emitted photons with energies distinct from the initial excitation. Atomic particles, like electrons, absorb light energy or photons and emit light radiation only in the form of electromagnetic waves called optical absorption. Thus, light, a form of electromagnetic radiation, can be absorbed by a sample material, converting its energy into internal energy.

8.7.2 Phosphorescence

Phosphorescence is known as luminescence, or also called delayed luminescence or "afterglow", that is, even though the excitation process is stopped, the emission of light becomes continuous. A phosphorescent material glows for a few minutes until the excited atoms in the phosphors spontaneously relax back to their ground state. Generally, phosphorescence emission in fluid solutions is usually masked at room temperature due to dominant non-radiative processes and quenching mechanisms. Phosphorescent substances are generally known as "glow-in-the-dark" materials. These materials are utilized to caution and alert drivers, enhancing road safety. These materials are employed for attention-grabbing advertisements, campaigns, and safety applications, warning people of potential risks. The toys used for children are generally made with luminescent substances and naturally glow because the material initially receives energy from light and emits light energy again.

The characteristics of the host material play a crucial role in determining the optical and spectroscopic studies of RE ions. To get practical applications in industries, the RE ion-doped glasses require hosts with superior chemical durability, thermal resistance, low melting temperature, and elevated refractive index. The properties of Ln^{3+} ion-doped glasses mainly depend on the host glass material, and in some cases, even a small change in the concentration of Ln^{3+} ions makes a lot of spectroscopic changes. The Ln^{3+} ion-doped glass materials are of great importance for their unusual optical properties [8–15].

8.8 Applications of Rare Earth Materials

8.8.1 Phosphors in Photonics

Phosphors are solid materials that display luminescence when exposed to radiation, including ultraviolet (UV) light or electron beams. The luminescence studies of the sulfide phosphors have been started after 1950 extensively. The first phosphor material is the barium sulfide which was prepared at low luminescence efficiency, and too many faults were identified due to fast decomposition in moist air. By using some metallic impurities and RE ions, the alkali earth sulfide phosphors are activated for practical importance. Good fluorescence has been obtained with RE ion-doped crystalline materials. In crystalline phosphors, the luminescence originates from the impurity system called the activator. Generally, the activator ions form luminescent centers where these ions are surrounded by host crystal ions. As a result, the excitation and emission can be observed in the phosphor material. Some co-activators are added to the system to enhance luminescence without significantly altering the emission spectral profile.

The use of phosphor materials is vital in display devices like CRTs, LCDs, LEDs, and plasma panels for efficient image production [34]. The display color on the screen depends on the lanthanide elements. The europium and yttrium combination gives a red color, the combination of terbium, fluoride, and zinc sulfide produces a green emission, while cerium, strontium, and sulfide compounds yield a blue color. These phosphors release light radiation known as luminescence when they are activated by photons [35].

8.8.2 Rare Earth Doped Glasses

The RE ion-doped glasses are potential materials for optical fiber amplifiers, optical detectors, laser-emitting devices, waveguide lasers, and fluorescent display devices [5–7]. Laser glasses have a lot of advantages over crystals because they can be prepared easily, including simplified manufacturing, flexible shaping, and improved optical homogeneity. Compared to crystals, glass materials can accommodate high concentrations of Ln^{3+} ions. Among all lanthanides, cerium has commercial applications and can be used as a decolorizing agent in glasses. When iron oxide is present in glass, it causes yellow and green colors, but by adding cerium, which oxidizes the iron [35], a small quantity of cerium will decolorize the glass into yellow and brown. In the same way, neodymium colors the glass red, praseodymium colors the glass green, holmium colors the glass blue, and erbium is used to color the glass pink [23].

8.8.3 Ceramics

Rare earth element oxides are essential components in ceramic material preparation to improve strength and toughness. Yttrium oxide and cerium oxides are used to minimize sintering temperature and also the production cost of ceramics. Rare earth oxides are also used in practical ceramics such as piezoelectric ceramics, semiconductor sensors, and microwave dielectrics [36]. To get the stabilized dielectric constant and to extend the life span of ceramic capacitors, the RE elements (La, Ce, Pr, and Nd) were added to the host materials [37].

8.8.4 Magnetic Materials

RE elements are used in the compounds of permanent magnets. A magnet was developed by alloying samarium with cobalt in the 1960s. Later, neodymium-iron-boron (NdFeB) magnets, introduced in the 1980s, have been largely replaced by samarium-cobalt (SmCo) magnets, except where high-temperature performance is required [38]. The magnetic energy of NdFeB magnets is about 2.5 times greater than that of SmCo magnets [23]. The neodymium magnets in smaller sizes are used in speakers of earphones, MP3 players, and other high-performance speakers. Some permanent magnets are used in DVD drives and hard disks [39]. Neodymium and iron with boron magnets have great importance in many environmentally friendly applications, particularly in green carbon-minimizing technologies, especially in wind turbines. For one-megawatt power production, neodymium magnets of nearly 0.6 to 1.0 tons are used, with almost 30% being RE elements [40, 41]. Using rare earth element magnets in air conditioning systems can reduce power consumption by 50% without changing its performance [42].

References

1. Connelly, N.G., T.D., Hartshorn, R.M., Hutton, A.T. (Eds.), 2005, *Cambridge: Royal Society of Chemistry publ.* ISBN0-85404-438-8. Archived from the original on 2008-05-27, Cambridge, Retrieved 2012-03-13.
2. Szabadvary, F., *Hand book on the Physics and Chemistry of Rare Earths*, vol. 11, K. Gschneidner and L. Eyring (Eds.), pp. 33–80, Elsevier, Amsterdam, 1998.
3. Haley, T., *Hand book on the Physics and Chemistry of Rare Earth*, vol. 4, K. Gschneidner and L. Eyring (Eds.), pp. 553–586, North-Holland Publishing Company, 1979.
4. Nandi, P., Srinivasan, A., Jose, G., *Opt. Mater.*, 31, 653, 2009.
5. Kamniskii, A.A., *Laser crystals*, second, Springer, Berlin, 1990.
6. Pecoraro, E., Sampaio, J.A., Nunes, L.A.O., Gama, S., Baesso, M.L., Spectroscopic properties of water free Nd_2O_3-doped low silica calcium aluminosilicate glasses, *J. Non-Cryst. Solids*, 277, 73, 2000.
7. Karthikeyan, B., Philips, R., Mohan, S., Optical and non-linear optical properties of Nd^{3+}-doped heavy metal borate glasses, *Opt. Commun.*, 246, 153, 2005.
8. Campbell, J.H., Suratwala, T.I., Nd-Doped Phosphate Glasses for High-Energy/High-Peak-Power Lasers, *Non-Cryst*, J., *Solids*, 263, 318, 2000.
9. Ehrmann, P.R. and Campbell, J.H., *J. Am. Ceram. Soc.*, 85, 1061, 2002.
10. Yang, J., Dai, S., Zhou, Y., Wen, L., Hu, L., Hiang, Z., *J. Appl. Phys.*, 93, 977, 2003.

11. Nunzi Conti, G., Berneschi, S., Bettinelli, M., Breuci, M., Chen, B., Pelli, S., Speghini, A., Righini, G.C., Rare-earth doped tungsten tellurite glasses and waveguides: fabrication and characterization, *J. Non-Cryst. Solids*, 345, 343, 2004.
12. Yeh, D.C., Sibley, W.A., Schneider, I., Afzal, R.S., Agarwal, I., *Appl. Phys.*, 60, 1648, 1991.
13. Holsa, J., Jungner, H., Lastuzaari, M., Niittykoski, J., Persistent luminescence of Eu^{2+} doped alkaline earth aluminates, MAl_2O_4:Eu^{2+}. *J. Alloys. Compds.*, 323, 326, 2001.
14. Morey, G.W., *The Properties of Glass*, Reinhold, New York, 1945.
15. Zarzycki, J., Glasses and Amorphous Materials, in: *Materials Science and Technology*, vol. 9, R.W. Cahn, P. Haasen, E.J. Kramer (Eds.), Basel, New York, 1991.
16. Weber, M.J., Science and technology of laser glass, *J. Non-Cryst. Solids*, 123, 208, 1990.
17. Peacock, R.D., The Intensities of Lanthanid f—f Transitions, *Struct. Bond.*, 22, 83, 1975.
18. Surendra Babu, S., Babu, P., Jayasankar, C.K., Tröster, T., Sievers, W., Wortmann, G., Photoluminescence from the $^5D_0$ level of Eu^{3+} ions in a phosphate glass under pressure, *J. Phys.: Condens. Matter*, 18, 1927, 2006.
19. Weeks, M.E., Discovery of the Elements. 6th Edition, *J. Chem. Educ. Am. Chem. Soc.*, 667–699, 1968, Ch. 16.
20. Auer von Welsbach, C., Über die Erden des Gadolinits von Ytterby. Monatshefte für Chemie und verwandter Teile anderer Wissenschaften, vol. 4, pp. 630–642, 1883, (now Chem. Monthly).
21. Cotton, S., *Lanthanide and Actinide Chemistry*, John Wiley & Sons Ltd, UK, 2006.
22. Emsley, J., *Nature's Building blocks: An A-Z guide to the elements*, Oxford University Press, Oxford, England, 2001.
23. Gupta, C.K. and Krishnamurthy, N., *Extractive Metallurgy of Rare Earths*, p. 508, CRC Press, Boca Raton, Florida, USA, 2005.
24. Dieke, G.H., *Spectra and Energy Levels of Rare Earth Ions in Crystals*, H.M. Crosswhite and H. Crosswhile (Eds.), Wiley Interscience, New York, 1968.
25. Dieke, G.H. and Crosswhite, H.M., The Spectra of the Doubly and Triply Ionized Rare Earths. *Appl. Opt.*, 2, 675, 1963.
26. Ogasawara, K., Watanabe, S., Sakai, Y., Toyoshima, H., Ishii, T., Brik, M.G., Tanaka, I., Calculations of Complete $4f^n$ and $4f^{n-1}$ $5d^1$ Energy Level Schemes of Free Trivalent Rare-Earth Ions, *Jpn. J. Appl. Phys.*, 43, L611, 2004.
27. Platt, A.W.G., Group trends, in: *The rare earth elements—fundamentals and applications*, Atwood (Ed.), pp. 44–53, Wiley, Chichester, 2012.
28. Shannon, R.D., Revised effective ionic radii and systematic studies of interatomic distances in halides and chalcogenides. *Acta Cryst A*, 32, 751–767, 1976.

29. Henderson, P., The rare earth elements: introduction and review, in: *Rare earth minerals, chemistry, origin and ore deposits, The Mineralogical Society Series*, vol. 7, chapter 1, pp. 1–19, Chapman and Hall, London, 1996.
30. Jensen, J. and Mackintosh, A., *Rare Earth magnetism—structures and excitations*, p. 413, Clarendon Press, Oxford, 1991.
31. Shionoya, S. and Yen, W. (Eds.), *Phosphor Hand book*, CRC Press, Boca Raton, Florida, USA 1998.
32. Andres, J. and Chauvin, A.S., Lanthanides: luminescence applications, in: *The rare earth elements—fundamentals and applications*, D.A. Atwood (Ed.), pp. 35–152, Wiley, Chichester, 2012.
33. Feldmanm, C., Justel, T., Ronda, C.R., Schmidt, P.J., Inorganic Luminescent Materials: 100 Years of Research and Application, *Adv. Funct. Mater.*, 13, 511, 2003.
34. Great Western Minerals Group LTD (GWMG), *Deep Sands*, 2010, http://www.gwmg.ca/html/projects/deep-sands/index.cfm.
35. Jackson, W.D. and Christiansen, G., US Geological survey circular 930-N, International Strategic Minerals Inventory Summary Report - Rare earth oxides, 1993.
36. Guanming, Q., Xikun, L., Tai, Q., Haitu, Z., Honghau, Y., Ruiting, M., Application of Rare Earths in Advanced Ceramic Materials. *J. Rare Earths*, 25, 281–285, 2007.
37. Roskill, *The economics of rare earths and yttrium*, 6th edition, Roskill Information Services Limited, London, United Kingdom 1986.
38. Castor, S.B. and Hedrick, J.B., Rare Erath Elements, in: *Industrial Minerals and Rocks: Commodities, Markets, and Uses*, 7th, J.E. Kogel, N.C. Trivedi, J.M. Barker, S.T. Krukowski, (Eds.), p. 1568, SME, Littleton, Colorado, USA, 2006.
39. Lynas Corporation LTD., *What are rare earths?*, 2010, Cited January 2010, www.lynascorp.com.
40. Avalon Rare Metals INC., Rare Earths 10, Rare Earth Elements And The Green Energy Economy, 2009, Cited February 2010, www.avalonraremetals.com.
41. Lifton, J., Is the rare earth supply crisis due to peak production capability or capacity?, 2009a, GLG News, 06/09/2009, www.glgroup.com.
42. London, I.P., Are we going to meet our growing need for rare earth supply? Avalon rare metals Inc. *Magnetics Conference 2009*, 2009.

9

Emerging Materials for Future Energy Storage and Energy Conversion Application

Sancheti Santosh D.[1*], Chandore Rahul N.[1], Vishal Jain[2] and Ingale Sachin P.[1]

[1]Department of Mechanical Engineering, SNJB's Late Sau. K. B. Jain College of Engineering, Chandwad, Maharashtra, India
[2]Department of Computer Science and Engineering, School of Engineering and Technology, Sharda University, Greater Noida, Uttar Pradesh, India

Abstract

The transformation from fossil fuels to clean energy is necessary as global energy demand increases. This chapter explores materials used for energy storage and conversion purposes. It also highlights the challenges of fossil fuels and the vital role of advanced materials for energy storage. This chapter explains key materials, including graphene, perovskites, and hydrogen, as well as their chemical properties and potential applications. Graphene provides high mechanical strength and conductivity, enhancing the battery's performance and efficiency. Similarly, hydrogen serves as a clean energy carrier. It also lightens thermoelectric materials and photovoltaic systems. By exploring these materials, this chapter aims to provide a comprehensive overview of the current scenario and future directions in energy storage and conversion. This chapter also gives detailed insights into different types of energy storage technologies like batteries, supercapacitors, fuel cells, etc. Advancements in battery materials can give higher specific energy and power density in real life.

Keywords: Energy storage systems, supercapacitors, graphene, electrochemical energy storage, renewable energy

**Corresponding author*: santusancheti@gmail.com

Kolla Bhanu Prakash, S. V. Ranganayakulu and K. S. Jagannatha Rao (eds.) New Frontiers in Materials Science: Interdisciplinary Approaches to Innovation and Technologies, (159–186) © 2025 Scrivener Publishing LLC

9.1 Importance of Energy Storage and Conversion in Modern Society

Energy consumption in modern society is continuously rising, which reflects critical global challenges in energy supply and climate protection. Fossil fuels are becoming more and more exhausted resources, and at the same time, they produce more and more greenhouse gases, which are closely related to global warming [1]. Therefore, it has become necessary to supply, store, and use clean energy sources. Renewable energy sources are one of the key solutions to overcome these challenges since they are porous, unlimited, and environmentally friendly. Current global energy demands are expected to rise by 50% until 2030, and it is predicted that renewable energy sources will take 30% of the energy supply by 2030 and 50% by 2050 [2]. The intermittency of energy supply from renewable energy sources, e.g., solar, wind, and tidal, is one of the main bottlenecks for widespread applications.

Finding solutions to rearrange the energy supply and demand is critical for the practical usage of renewable energy sources. It is therefore foreseen that energy storage will be a core solution to re-conciliate the decoupled energy supply and demand. Energy storage systems can be implemented at various levels, e.g., in buildings, factories, offices, neighborhoods, cities, and at the grid/station level. Energy storage technologies include mechanical, electrical, thermal, electrochemical, and chemical approaches. Energy storage systems have various scale and energy density options. Selecting and designing suitable energy storage technologies at different levels is an important topic. Energy storage has been a crucial part of energy demand and supply balance. The pumped-storage systems using water were employed decades ago to adjust the load and stability control of currently used electricity grids, and it has been a pivotal technology to ensure the grid's stable and controllable power supply.

9.1.1 Challenges Posed by Fossil Fuels and the Need for Clean Energy Alternatives

Despite being an effective energy source, fossil fuels pose various challenges, including environmental ones. Fossil fuels, including gas, coal, and diesel, are the primary sources of worldwide energy. They are heavily relied upon for heating, operating transportation, industries, and factories. The rising development in emerging economies and the increased use of fossil fuels have caused various environmental challenges, including significant increases in CO2 emissions. Thus, fossil fuel consumption must be significantly reduced to limit environmental issues [3]. The atmosphere's carbon sink has become saturated, increasing CO2 concentrations and global warming.

Additionally, fossil fuel reserves, fed by past solar energy, are gradually depleting. Oil will be the first fossil fuel to dwindle, ultimately leading to an energy crisis that will significantly affect modern human society; in approximately 30–50 years, fossil fuel extraction will cease. Accordingly, it is planned to create new, clean, and renewable energy sources to sustainably replace fossil fuels. For long-term energy supplies, energy sources like solar, wind, and geothermal are favorably considered [2].

9.1.2 Global Energy Demands and Renewable Energy Sources

All forms of energy are essential for human civilization. Energy is the lifeblood of industry, transport, and modern society. It is necessary to ensure food, water, and energy security and improve quality of life. However, energy has different availability and access. Worldwide, energy demands are steadily increasing [4]. This need is projected to increase to 405×10^{20} J by 2035 as the human population reaches 9 billion.

Currently, fossil fuels provide the majority of energy used on the Earth. However, fossil fuels are non-renewable energy sources and will eventually run out [5]. It is estimated that, at current consumption rates, oil could run out by 2065, natural gas by 2085, and coal by 2140. Burning these fuels to provide energy also produces greenhouse gases, which have severe consequences for global warming and have been implicated in the current climate crisis. Thus, the need for alternative forms of energy is becoming critical. Renewable energy sources could meet humanity's energy requirements. Figure 9.1 shows the transition from fossil fuel-based energy to renewable energy, highlighting its significance in addressing climate change and pollution issues.

9.1.3 Role of Energy Storage in Balancing Supply and Demand

Modern energy systems face the challenge of matching energy supply to demand. Balance is often defined as a state of equilibrium among contrasting forces or influences. The balance of an energy system must be considered in different time frames. On the shortest time scale, in the order

Figure 9.1 Transitioning from fossil fuel-based energy to renewable energy is crucial for addressing climate change and pollution issues.

of minutes, it refers to the instantaneous balance between total energy supply and that demanded by consumers; thus, maintaining a respective balance between energy generation and consumption across global power and energy grids requires the provision of either lossless storage or a controlled load that can absorb excess supply when needed [6]. This practical exercise shows that the smallest energy storage system with lossless hydro storage for an arbitrary distribution of hourly generation and consumption would need to have a storage volume of at least the size of the annual average generation to allow for long-term balance. In a time frame of seconds, primarily referring to significant power generation loss, the balance of an energy system is likewise necessary, thereby relying on automated control loops that support instantaneous protection.

On a broader footprint and in a longer time frame—of hours, days, months, and even years—balancing means providing the energy supply for the energy demanded by the consumers in a broader sense [7]. Here, larger volumes of energy must be either converted and stored or provided by a controlled load. For long-term energy balancing in a system with variable supply and controlled load, there is either a demand for a lossless storage system with a capacity that would have to equal or exceed the demand of the load at its maximum level or there must be a basis for long-term alternative generation such as hydraulic, geothermal, or solar forced flow generation, similar to infrastructure needs in a world without storage.

9.2 Role of Materials in Energy Storage and Conversion

Innovation of novel materials has been a critical driving force in energy storage and conversion. Energy storage systems, such as batteries and supercapacitors, are considered the heart of smart grids and electric vehicles, surfacing endless needs to invent advanced materials and their innovative applications [2]. Nevertheless, the development of materials for new energy conversion devices, such as organic photovoltaics and fuel cells, is also essential to utilize rechargeable sources of energy, such as solar, wind, and hydroelectric powers, surging ever more interest in materials development for a wide range of energy systems [8]. Figure 9.2 illustrates the key trends in materials research for energy storage and conversion systems, focusing on the trade-off between cost, performance, and stability.

Energy conversion is defined as the process of changing one form of energy into a different form. Various forms of energy can be converted into electrical energy. Though energy conversion techniques differ by energy source and utility, those techniques share a central role of materials. Materials with unique electronic structures and chemical properties play critical roles in energy conversion processes at atomic, molecular, and nano scales. Consequently, developing novel materials with unique and desired intrinsic characteristics has been a core driving force in the progress of energy conversion devices.

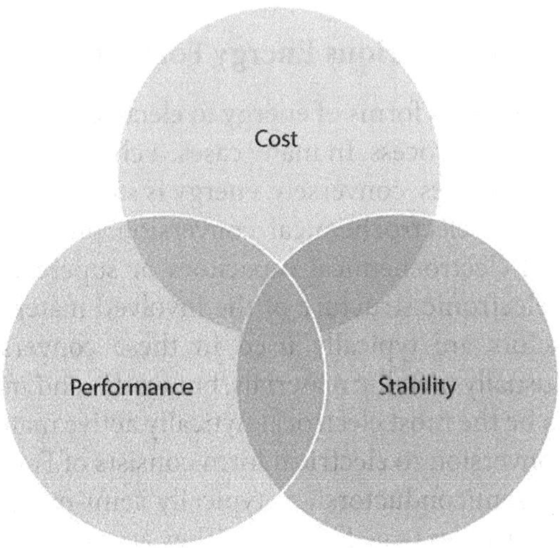

Figure 9.2 Trends in materials research for energy systems and devices.

9.2.1 The Importance of Material Innovation in Energy Systems

Material innovation will be central to efforts to realize and expand promising energy storage and conversion technologies. Many of these technologies rely on innovative materials, or materials in innovative configurations, to enhance device performance or efficiency. For example, the nanostructuring of the active components in composite electrodes can promote ion and electron transport beneficial for batteries or electrochemical capacitors, and it has led to immense improvements in specific power, rate-capability, and energy density in these technologies [2]. Nanostructured and well-ordered hierarchical porous electrodes can significantly enhance the activity and selectivity of catalyst materials in fuel cells, electrolyzers, and approaches for methane reforming and CO2 and organic waste valorization [9]. This ambition to innovate new materials, or materials in innovative configurations, will be paramount in advancing energy storage, conversion, or harvest technologies and propelling them from a nascent state to wide-scale commercial employment.

Fine control over material composition, morphology, structure, and functional properties of the active components in energy storage and conversion devices will be sought. This requires the development of new materials or new "chemical" formulations of existing materials that facilitate better control over microstructure at the nanoscale and the engineering of new material properties that enhance the performance and efficiency of energy storage and conversion technologies [33].

9.2.2 Conversion of Various Energy Forms to Electrical Energy

The conversion of various forms of energy to electrical energy is a straightforward yet significant process. In many cases, a chemical reaction releases energy in ample quantities; conversely, energy is stored in chemical bonds. The vast majority of electrochemical conversion processes, e.g., batteries, fuel cells, and electrochemical capacitors or supercapacitors, exploit a change in the electronic structure of the involved material. Conductors and semiconductors are typically used in these conversion processes. Conductors are usually metallic materials, but oxides and metallic carbides or metals tend to be the most electrocatalytically active materials [2]. Their mechanism of conversion to electrical form consists of Faradaic reactions. On the contrary, semiconductors are typically semi-metal oxides (iv) or semi-conducting carbides (e.g., SiC, TiC). They act as sensor materials and rely on the change of the solid-state conductivity to convert the surrounding energy to a variation in voltage [10].

9.3 Energy Storage Materials

Among the many materials for energy storage, graphene, perovskite, and hydrogen storage materials are essential. The current scenario of milestones achieved in electrochemical energy storage using carbon polymer nanocomposites and the technology transfer made possible through graphene and fullerene polymer nanocomposites was reviewed [11]. A comparison of specific capacitance, power density, and cyclic stability of different classes of carbon polymer nanocomposite supercapacitors reveals that they have attained higher performance than the best-commercialized systems available, thus indicating the bright prospects of this technology. The issues that need further attention in synthesizing carbon polymer nanocomposites for energy storage were pointed out concisely. A deeper understanding of the surface chemistry of polymer nanocomposites, which govern the electrochemical behavior, will permit significant performance increases, and low-cost, commercially viable systems will emerge.

With advancements in energy production, the interest in its storage is increasing, and so is the quest for novel materials for conversion. The experimental realization of many candidate materials at the lab scale for energy conversion stages has brought attention to their integration for whole storage unit configuration. The recent status of material research concerning energy storage devices, currently focusing on battery systems, was discussed [2]. Current trends in improving the working capability of a new class of materials called perovskites were presented. There is an increasing focus on metal hydrides as solid-state hydrogen storage material, issues dealing with the containment of physical gas storage systems, and the path forward for industries investing in advanced low-cost bulk alternative systems for Li-ion battery technology were highlighted.

9.3.1 Trends in Materials Research: Graphene, Perovskites, and Hydrogen Storage Materials

The emerging trends in materials research will be deliberated with a focus on materials for energy storage and conversion. Due to the over-consumption of fossil fuels, the need for sustainable energy sources like solar power is increasing. Because of the intermittent nature of solar power, there is an increasing demand for materials that can efficiently store energy for clean energy conversion. To match the increasing energy demands, new materials possessing better efficiency and longer life can be a better choice for energy conversion and storage applications. There is a significant need for sustainable and low-cost materials for electrochemical storage and conversion devices. Various organic, inorganic, and hybrid nanostructure-based materials are being investigated for electrochemical storage and conversion. Some of the exciting materials that can ignite interesting phenomena in the field of energy storage and conversion research, as per literature and the researchers' understanding, include:

1. **Graphene:** Since it emerged from isolation in 2004, graphene has attracted the major attention of material researchers owing to its fascinating combination of properties. It possesses the highest thermal and electrical conductivity, a high surface area (2630 m2 g−1), high mechanical strength and stability, excellent electron mobility, and a tunable band gap (0.016–3 eV). Owing to these unique properties, graphene has been explored for various applications such as in electronics, ultracapacitors/supercapacitors, lithium/sodium batteries, solar cells, electrocatalysts, drug delivery, bio-sensors, and anti-corrosion coatings, and as a potential candidate for hydrogen storage [11, 12].
2. **Perovskites:** The design of hybrid organic-inorganic perovskites (HOPs) has become a hot topic in the area of solar energy conversion, especially visible-light-driven solar energy harvesting and storage. As a new member of the sunlight harvesting family, HOPs possess several useful characteristics, such as a mixture of organics and strong inorganic covalent bonds, which provide a tunable band gap (about 1.5–3.0 eV) and multi-channel harvesting of the entire visible spectrum. HOPs can inspire new designs of metal-organic nanoclusters, which not only crystallize within metal nanoparticles but also retain their remarkable stable electronic properties.
3. **Hydrogen Storage Materials:** Hydrogen storage is vital for transitioning from a fossil-fuel-based economy to one where hydrogen is used as an energy carrier. Metallic hydrides have been the materials of choice for portable and stationary hydrogen energy systems for decades. Specifically, metal

hydride construction technology has been applied to hydrogen storage tanks and reactors for hydrogen energy materials for motives such as underground storage. Nanostructured metal hydrides exhibit a number of advantages, most notably a significant reduction in hydrogenation and dehydrogenation temperatures. Currently, research is focused on more advanced hydrogen storage materials based on complex metal amides and complex hydrides of light metals to store hydrogen on-board light-duty vehicles at 70 bar under a hydrogen absorption/desorption cycle of hydrogen gas.

9.3.2 Overview of Different Storage Technologies

Energy storage technologies are not unique; they can be categorized into different umbrellas, each depicting different storage approaches: mechanical, electrochemical, thermal, chemical, electromagnetic, and pumped hydroenergy storage. These technologies each have their characteristics and are used differently to tackle energy storage challenges in the world [7]. Over the last decades, with the comprehensive expansion of electrical system technologies, especially smart grid technologies, the diversification of energy resources has brought forth unprecedented challenges in energy management, one of which is energy storage. Energy storage refers to a process whereby energies in different forms can be stored and released later for various purposes. It is understood that energy storage can significantly improve the overall performance of modern electrical systems such as power systems and smart grids [13].

9.4 Batteries

Batteries are widely used electrochemical power sources that convert chemical energy to electrical energy, or vice versa, through redox reactions.

Batteries can be classified into:

(1) Primary batteries and
(2) Secondary batteries (rechargeable batteries) [14].

Batteries that can be used only once are defined as primary batteries. The reaction is irreversible, and the consumed materials are not restored.

Batteries that can be discharged and recharged many times are defined as secondary batteries. The electrode materials (anode and cathode) are restored during charging and discharging, making the energy conversion cycle repeatable.

Secondary batteries can be divided into:

(1) Liquid-state batteries
(2) Solid-state batteries,
(3) Metal-state batteries, and
(4) Flexible batteries.

Liquid-state batteries contain at least a liquid component, making them chemically stable, and they will be the first to be commercialized. Due to their potential benefits, solid-state and flexible batteries are being actively researched and developed. Besides, the current state of sodium-ion batteries, redox flow batteries, and various metal batteries such as lithium, potassium, aluminum, and magnesium batteries are introduced [15].

Batteries can also be categorized into:

1. Alkaline batteries
2. Lead-acid batteries
3. Lithium-ion batteries,
4. Nickel-metal hydride batteries, and
5. Nickel-cadmium batteries.

Alkaline batteries are primary batteries. They usually use zinc powder as the anode material and manganese dioxide as the cathode material. Alkaline batteries have advantages such as long shelf life, high energy density, and environmental safety. Lead-acid batteries contain lead dioxide and sponge lead as the cathode and anode materials, respectively. It is a mature and well-established battery technology. The current usage of these batteries is in cars, e-bikes, and stationary systems. This battery technology is often used for energy storage connected to renewable energy power generation systems. Lithium-ion batteries consist of transition metal oxides as the cathode and graphite as the anode material. Lithium-ion batteries mainly power consumer electronics, electric vehicles, and electric energy storage systems. Nickel-metal hydride batteries consist of nickel hydroxide as the cathode material and metal hydride as the anode material. It is also a mature battery technology, mainly used in hybrid electric vehicles.

9.4.1 Types of Batteries: Alkaline, Lead-Acid, Lithium-Ion, and Nickel-Metal Hydride

9.4.1.1 Alkaline Batteries

Alkaline batteries are among the most commonly used types of batteries. Like zinc-carbon batteries, they are non-rechargeable batteries that use zinc as the anode (negative terminal) and either manganese oxide or nickel oxide as the cathode (positive terminal). Alkaline batteries are marketed as having a longer life than zinc-carbon batteries. They offer a higher energy density, operating life, and better time consistency than previous formulas—at about twice the price.

9.4.1.2 Lead-Acid Batteries

Lead-acid batteries are rechargeable batteries that use lead as the anode and lead dioxide as the cathode. As the name implies, sulfuric acid is the electrolyte medium for conduction. Lead-acid batteries are well-known for being substantial (heavy and bulky) batteries primarily used for energy storage. The lead-acid battery was invented in 1859 and is one of the oldest single-cell energy storage systems. Because of this, it is the benchmark for comparing newer battery technologies. Lead-acid batteries have an efficiency of 90%, can be recharged in 8–16 hours, and retain about 60% of their charge for 4 to 6 months without use. They are well-suited for tasks requiring high capacity demand over short usage distances.

Nonetheless, lead-acid batteries have considerable drawbacks. They are heavy batteries (for optimal energy density) that are not fully sealed, allowing corrosive electrolytes to leak out. Lead-acid batteries can degrade rapidly (e.g., in as little as a day) when left uncharged for a long time and discharged to low voltages. Sulfation occurs when lead sulfate crystals build up on the electrodes of lead-acid batteries from being discharged too deeply or remaining in a discharged state. This short-circuits the battery (freezing the cell's voltage) and significantly decreases capacity (hence how long a battery lasts). Sulfation can occur within a month of being discharged.

9.4.1.3 Lithium-Ion Batteries

Lithium-ion batteries are rechargeable batteries widely used in cell phones, laptops, cameras, and electric vehicles. They are made up of lithium cobalt oxide as the cathode, graphite as the anode, and lithium salt as the electrolyte. Their most defining factor is their relatively high energy density of 200-300 Wh/kg, meaning they can store a considerable quantity of energy in a small volume. However, lithium-ion batteries contain organic solvents that need to remain unexposed to the surrounding environment. Lithium-ion batteries also degrade more rapidly when exposed to cold temperatures or higher internal temperatures, resulting in their usage being confined to consumer electronics. Nevertheless, massive infrastructure development has been made to accommodate future usage of lithium-ion batteries in electric vehicles. Figure 9.3 provides a schematic representation of the working mechanism of a lithium-ion battery, depicting ion movement between the cathode and anode during charge and discharge cycles.

9.4.1.4 Nickel-Metal Hydride Batteries

Nickel-metal hydride batteries are rechargeable batteries that store energy using a nickel oxide-hydroxide cathode, an alloy anode, and an aqueous potassium hydroxide electrolyte. They are environmentally friendly and cost-effective alternatives to both zinc-cadmium batteries and lithium-ion batteries. Nickel-metal hydride batteries feature greater volumetric energy density than lithium-cobalt batteries due to their high-capacity electrodes and relatively higher operating voltage. The more recent development of complex metal hydrides serving as the alloy anode allows very compact

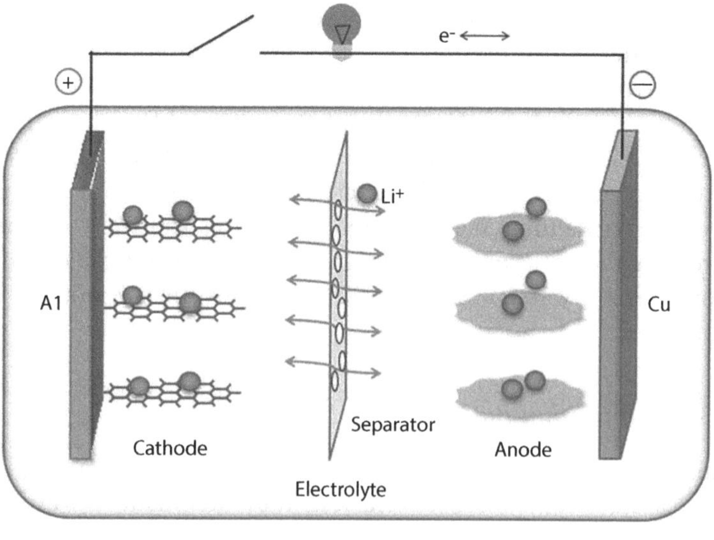

Figure 9.3 Schematic diagram of the working mechanism of lithium-ion battery.

battery systems for electric vehicle applications. Overall, nickel-metal hydride batteries are relatively novel batteries that maximize energy density while preventing environmental hazards from toxic marine life [16].

9.4.2 Advantages and Limitations of Each Type

Alkaline batteries are commonly used in household electronics, toys, and appliances. They have four times the energy density of zinc-carbon batteries and operate in a wider temperature range [17]. At room temperature, the battery voltage is about 1.5 volts. They contain a gelled zinc anode and a powdered manganese dioxide cathode mixed with potassium hydroxide electrolyte, which is safe and enables easy disposal. However, alkaline batteries cannot be recharged and cannot withstand freeze-thaw cycles, although there are some rechargeable varieties available [18].

Lead-acid batteries are commonly used in automobiles, uninterruptible power supplies, and other large systems. They have a low initial cost and high discharge rate capability, thanks to the high density of the negative electrode. They can also be started in cold weather due to the small temperature coefficient of the discharge reaction. However, they have a low energy density, high self-discharge rate, and cycle life. Additionally, they are highly toxic. Improvements include the use of gelled, non-spillable systems.

Lithium-ion batteries are the most widely used rechargeable batteries in portable electronics. They have a high energy density and are safe to use, but large cells can be hazardous unless carefully designed and protected. High-power Li-ion cells are often made using lithium iron phosphate (LiFePO4) cathodes, which have a lower energy density and higher manufacturing cost due to the need for casing. Energy density cells have a nickel-cobalt-aluminate oxide cathode, which is cheaper but has more temperature-sensitive materials. There are more exotic Li-ion cells available, such as the deployment of LiNi0.5Co0.2Mn0.3O2 for the Tesla Model S. Nickel-metal hydride batteries are used in hybrid vehicles and have a much higher energy density than nickel-cadmium batteries. However, they have a higher cost and poor cycling performance at low temperatures.

9.5 Supercapacitors

Among energy storage systems, there is no single technology that can completely resolve the dilemma of energy storage and conversion. Batteries and supercapacitors are the two most widely utilized electrochemical energy storage devices. Focus is on supercapacitors, which is especially attractive because of their ability to take advantage of ionic charge movement between electrodes and electrolytes, which is fundamentally different from the mechanisms associated with electrodes in dynamic conversion devices. Thus, the timescale for charging and discharging and the power density of supercapacitors are much higher than batteries. On the other hand, supercapacitors' energy density remains lower than batteries because the energy stored in electrochemical devices is proportional to operating voltages, which are lower for double-layer capacitance and redox-type faradaic capacitance. Figure 9.4 presents a schematic diagram of a supercapacitor in a charged state, showing how electrolyte ions interact with porous electrodes to store energy.

There are many factors that can affect the performance of supercapacitors including electrode materials and designs, electrolytes, separators, current collector properties, and supercapacitor interface configurations. Among these materials, transition metal oxides/sulfides, conductive polymers, and carbonaceous materials are being utilized for a variety of supercapacitor applications. Mainly because of the relatively low conductivity of metal oxide/sulfide materials, which are usually in nanosized or microsized form to minimize ionic diffusion distances and to maximize specific surface areas at elevated charge delivery rates [19, 34], fabrication of supercapacitors based on this type of pseudocapacitive material is mainly limited to the use of film-type or paste-type structures built on metal foil current collectors. This essentially requires complicated and often costly fabrication processes and limits the designs of the electrochemical devices.

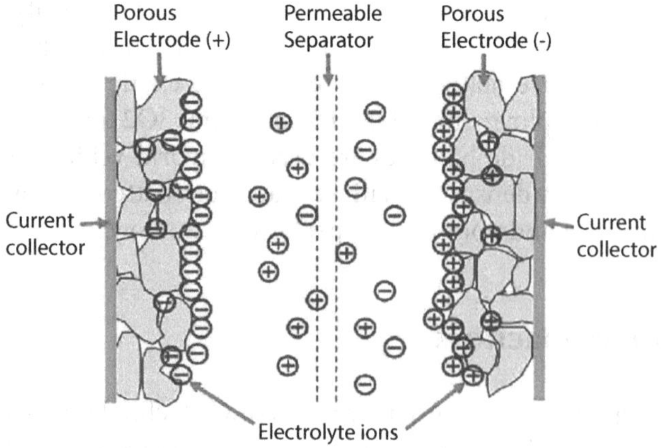

Figure 9.4 A schematic diagram of a supercapacitor at the charged state.

9.5.1 Structure and Function

In recent years, the research on energy storage and conversion devices has gained prime interest all over the world especially due to the energy crisis and rapid industrialization. The unit of energy storage is often expressed in farads (F). This unit was named in honor of Michael Faraday, who made significant contributions to the field of electromagnetism and energy storage. Generally, supercapacitors are counted as higher Farad devices. Supercapacitors are electrochemical energy storage devices that exhibit high energy and power density, high safety, long life cycles, and low maintenance. Supercapacitors are familiar as supercapatteries or hybrid capacitors due to the combination of capacitor and battery mechanisms. These terms are widely used in various publications. They operate on either double-layer capacitors or electrochemical pseudocapacitance through interfacial charge separation and concentration variation, whereas batteries operate on faradaic or redox reactions through electronic transfer between electroactive materials [19, 20].

Due to their unique characteristics, supercapacitors are entirely different from other traditional batteries and capacitors. Depending on their properties, supercapacitors can solve the specifications of individual applications. Flexibility in the combination of materials for electrodes, electrolytes, and separators with hybrid configurations enhances the chance of material selection in supercapacitors. Undoubtedly, they can be pointed out as the best energy storage system to meet even the most demanding requirements. The rapid advancement and performance of supercapacitors compete with rechargeable batteries. Various research and development efforts are being conducted to keep growing the market and commercial values of this energy storage device. It plays a significant role in electrical and electronic devices (such as mobile phones, laptops, electric vehicles, and many more) with high performance and affordable prices.

9.5.2 Advantages Over Traditional Batteries

Supercapacitors, also known as ultracapacitors or electrochemical capacitors, are either fluid or solid-state devices that store electrical energy by forming an electrostatic double layer at the electrode surfaces. This mechanism is similar to that of traditional capacitors. However, supercapacitors have a higher operating potential due to the pseudocapacitance provided by redox-active species. They bridge the functional gap between conventional capacitors and batteries. Although the power output of supercapacitors is lower than that of electrolytic capacitors, their specific energy is significantly higher. Batteries, on the other hand, can store more energy but at the expense of lower power densities [21]. While capacitors discharge within seconds, batteries take tens to hundreds of seconds. Supercapacitors can be charged quickly, often in under an hour, and have a remarkable cycle life, ranging from 100,000 to 1 million cycles. This longevity is due to their ability to store charges physically at the electrode surfaces without undergoing irreversible chemical reactions, which also results in less heat generation compared to batteries [19]. Additionally, supercapacitors can perform well at temperatures as low as −40°C, a feat not typically achievable by batteries, and are considered safer for high-power applications during charging and discharging.

9.6 Fuel Cells

Fuel cells are systems that convert the chemical energy of a fuel into electrical energy through electrochemical processes. This energy conversion can be written in the form of fuel oxidation and reduction reactions. The most common fuel for fuel cells is hydrogen, and oxidants are typically oxygen or air. Fuel cells are divided according to their types of electrolyte and proton conduction, such as phosphoric acid fuel cells (PAFC), proton exchange membrane fuel cells (PEMFC or PEF), alkaline fuel cells (AFC), and molten carbonate fuel cells (MCFC). Each type of fuel cell has its own characteristic temperature and pressure bands for normal operation.

Fuel cells can also be divided according to their types of reactants. For instance, nickel fuel cells (NiFC) and direct methanol fuel cells (DMFC) utilize gaseous or liquid forms of hydrocarbon fuels, respectively. Solid oxide fuel cell (SOFCs) are unique among fuel cells in that they use solid-state electrolytes that conduct oxygen ions instead of protons. SOFCs operate around temperatures ranging from 600–1,000°C and utilize ceramic electrolytes, such as ZrO2, stabilized with rare-earth metal oxides like Y2O3 or Sm2O3.

Fuel cell applications can be diversified into portable and fixed categories. Portable fuel cells, which require a high-energy-density and lightweight design, are used in applications such as laptop or mobile phone backup power sources. Similar power supplies with larger scales are fit for fuel cell appliances in the products of domestic consumer electronics. Small-sized fuel cells are portable power sources for small electric vehicles such as toy cars, motorbikes, and bicycles. SOFCs, characterized by negligible space requirements structures, are used for medium-capacity stationary electric power delivery and can be engine-mounted auxiliary power sources. Due to the nature of the operation, fuel cells can be launched in a parallel battery configuration, providing an uninterruptible supply of energy that is hardly likely to fail. The battery discharges energy accumulated as long as needed. At the same time, the fuel cells continuously produce energy from hydrogen fuel stoichiometric feed to ensure the same output power as drained by an electric load. Fuel cell technology is now targeting the delivery of power in a newly growing category of vehicles and appliances—fuel-cell-powered ones. More dedicated applications of fuel cells seek optimum energy conservation, like transportation or stand-alone power and heat generation in domestic, commercial, and industrial sectors, where fuel cell by-products can directly substitute for energy sources at the site.

Fuel cells are typically composed of oxidant and reactant supply ports, electric terminals, a stack with multiple cells connected in series or parallel for increased output voltage or current density, contactors with distribution plates, separators preventing the mixing of fuel and oxidants, and a thermal control system to manage the voltage equipment temperature. The stack works under fuel and oxidant stoichiometric feeding for electrochemical reactions in electrodes at the electrodes/electrolyte interface. The ensuing processes away from the electrodes include reactants transported with turbulence mixing against the backpressure and through tortuous paths via porous distribution plates and flow channels. Temperature drops because of heat losses as the thermal energy of the mixture is converted into electrochemical activity plus Joule heat loss due to charge transport. A pressure drop along the flow channel leads to the emptying of the initially filled electrode pores as an output of water on reaction mass transportation. The water is formed on the anode side of the PEFC or the cathode side of the SOFC and separates from the gas mixture in gaseous form at the diffusion layer/flow field interface.

In contrast to the electrocatalytic reaction, no excess water is in the dry reactants supplied during operation. Thus, the conversion of the gas is disrupted away from the electrodes. Under PEFC conditions, extraction from the pores is faster than water adsorption, resulting in a two-phase gas and water flow. This latter means that a constant and nonzero outflow gas system is present at some point downstream of the flow channel.

9.6.1 Types and Applications

In the field of energy conversion, hydrogen's chemical energy can be transformed into electricity through solid oxide fuel cells, which store the generated electricity to be used as needed. Any surplus electricity is then used to produce hydrogen through solid oxide electrolysis cells in the energy reuse stage. The resulting hydrogen is then utilized again in the energy conversion stage, completing the energy circulation. This energy system consists of three parts: energy conversion, storage, and reuse, each containing devices made of different materials. Research is focused on understanding and developing these energy conversion, storage, and reuse devices. By comprehensively understanding and optimizing the materials used in these devices, valuable insight into the overall hydrogen energy system can be gained. Working with nanoscale materials poses challenges in terms of producing them in large quantities, as well as understanding and optimizing them.

Nonetheless, research aims to offer insight into the fabrication and analysis of nanomaterials tailored for each device. The solid oxide fuel cell (SOFC) carries out the most basic energy conversion process (fuel → electrical energy), boasting over 50% higher efficiency compared to traditional combustion engines.

Furthermore, due to its high operating temperature, the SOFC can be directly used with a variety of fuels, including hydrocarbons, without needing additional reforming equipment. These fuel cells are mechanically ideal as all their components are solid, and they operate quietly and with minimal emissions of pollutants, making them a highly efficient and durable potential system [9, 22].

9.6.2 Hydrogen Storage and Conversion

Hydrogen storage and conversion using fuel cell technology play a key role in the hydrogen energy process. Hydrogen can be produced from various renewable sources. It can be stored in a gas or liquid state. Hydrogen can be consumed in internal combustion engines or specially designed fuel cells. Different technologies exist for this purpose. Environmental concerns regarding CO2 emissions from fossil fuel energy sources, growing energy needs, and the use of renewable energy sources have led to worldwide interest in hydrogen as a clean energy carrier. Hydrogen energy has many benefits as a clean energy vector: it can be produced from various renewable energy sources such as wind, sun, geothermal energy, and biomass; it produces only water as a byproduct during the energy conversion process; and it can be utilized either by internal combustion engines or via fuel cells [23]. Figure 9.5 depicts the schematic structure of a hydrogen fuel cell, explaining the movement of electrons and hydrogen ions during energy conversion.

However, many challenges need to be solved for the hydrogen energy system to be fully implemented. Hydrogen storage has always been a key issue in the development of hydrogen energy. Mobile and fixed hydrogen storage systems need to be developed for the hydrogen energy economy. There will be special high purity requirements for hydrogen in certain applications. For example, polymer electrolyte membrane (PEM) fuel cells are expected to replace traditional internal combustion engines in a hydrogen energy economy, but hydrogen from fossil fuel reformers contains a lot of CO (up to 10,000 ppm) that could poison platinum catalysts. Gaseous hydrogen storage is a hydrogen storage method that uses a high-pressure vessel to store hydrogen gas at high pressure. High-pressure tanks are thick-wall cylinders with typical storage pressures of 200 bar. However, new tests storing 350 bar cylinders are being developed that are made of carbon fiber reinforced polymer (CF-RP) to save weight. There are hydrogen storage tanks of smaller size for homes used for power generation. These vessels are usually fabricated based on american society of mechanical engineers (ASME) standards. Material selection and design can be modified depending on the specific operating environments and requirements [24].

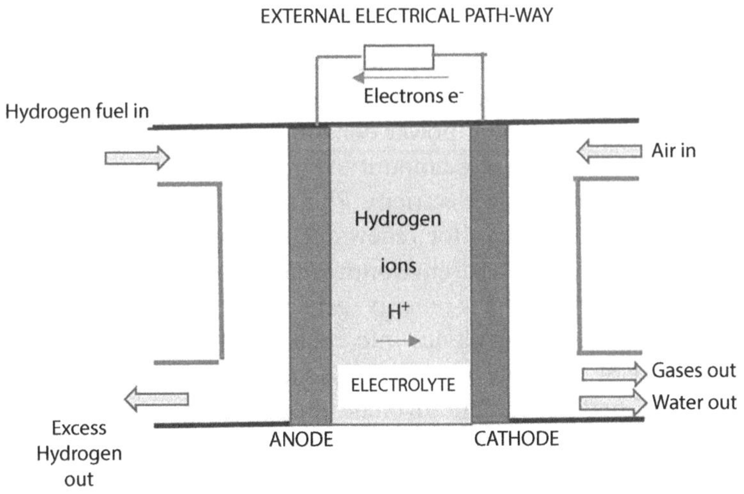

Figure 9.5 Schematic diagram of a hydrogen fuel cell.

9.7 Energy Conversion Materials

This special issue deals with energy conversion materials, which hold the primary function in systems that utilize such materials to drive periodic energy conversion processes. Energy conversion devices comprise a collection of materials, typically either catalysts or membranes, that are configured to convert one form of energy into another. Numerous devices exist, such as fuel cells and electrolyzers, solid-state batteries, photovoltaic systems, thermoelectric materials, and so on. Advances in energy conversion materials and devices have the potential to shape the energy systems of the future [2].

This special issue aims to cover a range of energy conversion materials, from metal oxide catalysts to organic nanomaterials, metal vacancies, and metals as well as their alloys, amidst various device configurations. It is hoped that this collection of materials will encourage and foster collaboration between scientists working in diverse fields of energy conversion processes [9].

9.7.1 Photovoltaic Systems

Photovoltaic systems, also known as solar photovoltaic systems, are energy conversion systems that generate electrical energy from solar radiation. Solar energy is the most abundant renewable energy resource since it surpasses worldwide customers' power demand by more than 5000 [25]. The Earth receives a considerable amount of solar radiation daily that can be harvested and converted to electricity. Photovoltaic systems are one of the most promising alternatives for renewable energy sources since they are quiet, low maintenance, and environmentally friendly due to no greenhouse gas emissions during energy generation [26]. Therefore, many countries integrate photovoltaics into their renewable energy initiatives to satisfy part of their power demand. Due to this demand and extensive research performed, photovoltaic energy conversion technologies have achieved substantial efficiency improvements and cost reductions. Additionally, the number of possible applications for photovoltaic systems has grown significantly as new architectures and form factors have emerged. Photovoltaic systems can operate on individual terrestrial rooftops mounted on buildings or vast areas of ground in parks (also known as grid-connected systems). They can even be mounted on water bodies, large windows, facade integration, etc.

The emergence of competitive stand-alone applications, from small consumer appliances powered by small solar panels to solar water pumps or telecommunication stations powered by large solar photovoltaic arrays, has also emerged. Photovoltaic systems can even be coupled with batteries or hydrogen production facilities to increase users' energy autonomy and avoid price fluctuations in the fuel energy market. When there is a need for larger quantities of energy, and the photovoltaic power plants become larger and larger (with several megawatts of installed power), the technology is no longer considered an energy user or complement but a competitor to traditional energy technologies. These predominant photovoltaic technologies in almost all proposed applications are based on silicon wafer solar cells. Although 40 years have passed since they were first commercialized, they are still the most used photovoltaic energy conversion technology, accounting for about 90% of worldwide annual photovoltaic power installations [35]. Figure 9.6 illustrates the working of photovoltaic systems, demonstrating how electrons move within the silicon layers to generate electric potential from solar energy.

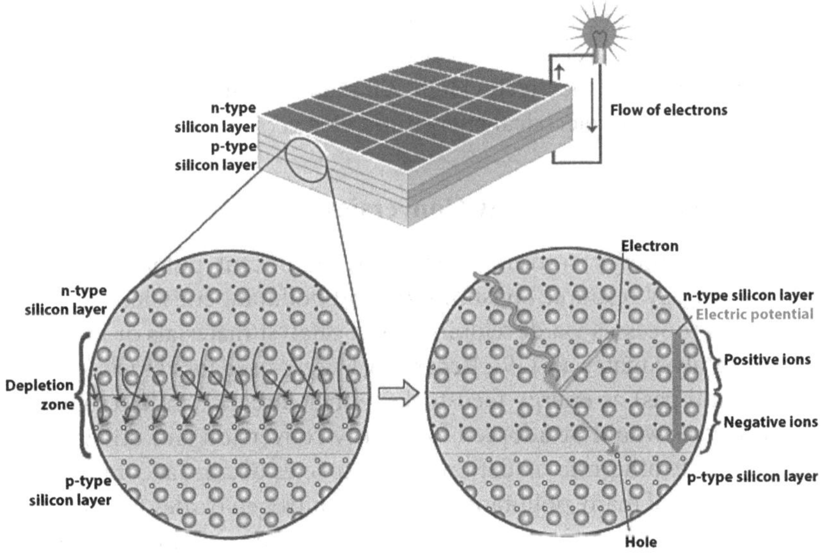

Figure 9.6 Photovoltaic systems.

9.7.2 Thermoelectric Materials

Despite advancements in electrical power generation and utilization, conversion efficiency remains low. It has been estimated that about 30% of generated electrical energy is dissipated as waste heat through thermal power generation and subsequent thermal management. In addition, heat dissipation generated by electronics, automotive, and other commercial devices imposes thermal waste. A promising approach to addressing these energy conversion challenges is temperature differential (TD) energy conversion. Heat can be converted into electrical energy if there is a difference in temperature. Thermoelectric materials could fulfill these requirements. They can generate electricity from temperature gradients through the Seebeck effect, where electrons diffuse from the hot region to the cold region, generating a potential difference [27].

In essence, a thermoelectric material has three unique properties: a high Seebeck coefficient (S), a high electrical conductivity (σ), and a low thermal conductivity (κ). The Seebeck coefficient indicates the tendency of a material to generate a potential difference in response to temperature gradients. The thermal conductivity indicates how well the material conducts heat. The combination of these transport properties gives rise to the dimensionless figure of merit (ZT) of a thermoelectric material: $ZT = (S^2\sigma)/(\kappa_e + \kappa_l)$, where κ_e and κ_l are the electronic and lattice contributions to thermal conductivity, respectively. High ZT materials are desired to convert thermal energy into electrical energy and vice versa [28].

9.7.3 Catalysts for Energy Conversion

Focusing on energy conversion systems, significant potential exists in reducing energy resource consumption and pollutant emissions of energy conversion processes via developing novel catalyst materials or improving existing catalyst materials [29]. Regarding widely-used fossil fuels (e.g., oils, coal, natural gas) as energy feedstocks, either direct or indirect combustion processes would be adopted for energy conversion and utilization, in which rapid exothermic oxidation of fuel species occurs, accompanied by the formation of many pollutant species like nitrogen oxides (NOx), sulfur oxides (SOx), and particulate matters (PM). It is essential to develop efficient technologies for fuel oxidation, such as NO decomposition and soot oxidation, to purify emissions. However, problems like energy resource over-consumption and long-term catalyst deactivation due to catalyst sintering and coke deposition plague the development of such technologies.

Out of this concern, the quest for novel functional materials and systems for energy conversion has achieved considerable momentum, in which molecule or ion species tend to be used as energy resources due to their natural abundance and low cost. In this regard, tremendous effort has been devoted to developing highly efficient and cost-effective catalyst materials in recent years for green energy conversion reactions, such as fuel cell (FC) reactions, CO2 reduction, and H2O splitting reactions [30].

9.8 Characterization Methods for Energy Materials

Characteristics of energy materials are analyzed through various methods termed characterization techniques. Several methods that are frequently employed to ascertain the internal structure include X-ray diffraction and neutron diffraction. Images of a material's surface or an object with resolutions up to even the atomic scale can be obtained using various microscopy techniques, an example of which is scanning electron microscopy. Nuclear magnetic resonance, X-ray photoelectron spectroscopy, Raman scattering, and UV-visible absorption spectroscopy can be effectively deployed to elucidate the compositional attributes of a solid material. Techniques are also available to study the performance of a material in an integrated fashion, such as thermogravimetric analysis for energy storage materials. Electrochemical impedance spectroscopy mainly explains the electrochemical performance concerning storage and conversion. Besides these techniques, several methods are employed for studying the interatomic distances at the inner shells of atoms, such as X-ray absorption spectroscopy, extended X-ray absorption fine structure, and X-ray emission spectroscopy [31].

Analyzing energy materials and determining their critical aspects of function, application, and effect is essential for developing these intelligent materials. In a way, energy materials represent functional materials with an additional dimension to their applications and effects. Energy materials are typically used for storage and conversion, which comprise materials' energy domain. The performance evaluation of energy materials critically demands characterization, i.e., deriving awareness of material aspects. To evaluate the versatility of these energy materials, sophisticated techniques such as *in-situ* and operando techniques are utilized to portray how these materials function in their actual operating conditions: the practical situation [32].

9.9 Conclusion

Sustainable and renewable energy sources have become urgent due to the current energy crisis and environmental pollution. Solar cells, fuel cells, hydrogen evolution electrodes, lithium-ion batteries, etc., have attracted substantial attention in modern harmonic life. However, the commercialization of these devices has been suppressed because of some issues, especially with energy storage devices. Therefore, it is essential to perform research on these materials, including synthesis, modification, structure, performance, and mechanism, to cover these indispensable social demands. Non-noble metal and noble hybrid materials exhibit apparent advantages in reducing the volume of materials and the costs used in the devices.

Furthermore, the cooperation of coupling processes between electrocatalysis and supercapacitors or lithium-ion batteries highlights the improvement in the performance of materials and devices. This review comprehensively emphasizes the recent progress in electrocatalytic activities of nanostructured hybrid materials and the mechanisms during the electrocatalytic process. Briefly reviewing the emphasis of the conclusion, this work provides readers with a relatively complete knowledge of the synthesis, modification, performance, and mechanisms of the essential hybrid materials for the electrochemical and energy fields.

References

1. Baumann, M., Historic and potential technology transition paths of grid battery storage: Co-evolution of energy grid, electric mobility and batteries, in: *Proceedings of the Engineering, Environmental Science*, 2015, https://api.semanticscholar.org/CorpusID:55938280.
2. Kim, J.K., Novel Materials for Sustainable Energy Conversion and Storage. *Materials*, 13, 11, 2475, 2020, https://doi.org/10.3390/ma13112475.
3. Aschbacher, L., Sancetta, B., Peterson, K., Degreenia, T., *21st century energy production & resulting impacts on society*, Worcester Polytechnic Institute, Worcester, Massachusetts, USA, 2012.
4. Schirone, L. and Pellitteri, F., Energy policies and sustainable management of energy sources. *Sustainability*, 9, 12, 2321, 2017, https://doi.org/10.3390/su9122321.
5. Yunus, A., Alharbi, Y., Abu-Siada, A., Sherkat Masoum, M., Overview of storage energy systems for renewable energy system application, in: *The 3rd Makassar International Conference on Electrical Engineering and Informatics*, 2012.

6. Götz, M., Lefebvre, J., Mörs, F., McDaniel Koch, A., Graf, F., Bajohr, S., Reimert, R., Kolb, T., Renewable power-to-gas: A technological and economic review. *Renew. Energy*, 85, 1371–1390, 2016, https://doi.org/10.1016/j.renene.2015.07.066.
7. Kung, C.-Y., Wang, T.-L., Lin, H.-Y., Yang, C.-H., A high-performance covalently bonded self-doped polyaniline–graphene assembly film with superior stability for supercapacitors. *J. Power Sources*, 490, 229538, 2021, https://doi.org/10.1016/j.jpowsour.2021.229538.
8. Wu, H.B. and Lou, X.W.D., Metal-organic frameworks and their derived materials for electrochemical energy storage and conversion: Promises and challenges. *Sci. Adv.*, *3*, 12, eaap9252, 2017, https://doi.org/10.1126/sciadv.aap9252.
9. Kim, S., Engineering the Morphology & Interface of Nano-sized metal oxides for Highly efficient Energy Devices, 2019.
10. Lukatskaya, M.R., Dunn, B., Gogotsi, Y., Multidimensional materials and device architectures for future hybrid energy storage. *Nat. Commun.*, *7*, 12647, 2016, https://doi.org/10.1038/ncomms12647.
11. Gopal, J., Muthu, M., Sivanesan, I., A comprehensive compilation of graphene/fullerenes polymer nanocomposites for electrochemical energy storage. *Polym. (Basel)*, *15*, 3, 701, 2023, https://doi.org/10.3390/polym15030701.
12. Hannan, M.A., Hoque, M.M., Mohamed, A., Ayob, A., Review of energy storage systems for electric vehicle applications: Issues and challenges. *Renew. Sustain. Energy Rev.*, 69, 771–789, 2017.
13. Korkmaz, E., Technically and Economically Viable Future Electricity and Fuel Storage Technologies, 2019.
14. Luo, R., Huang, Y., Wu, F., Li, L., Advanced high energy density secondary batteries with multi-electron reaction materials. *Adv. Science*, 31600051, 1–39, 2016, https://doi.org/10.1002/advs.201600051.
15. Nestler, T., Meutzner, F., Kabanov, A.A., Zschornak, M., Leisegang, T., Meyer, D.C., Combined theoretical approach for identifying battery materials: Al^{3+} mobility in oxides. *Chem. Mater.*, *31*, 3, 737–747, 2019, https://doi.org/10.1021/acs.chemmater.8b03631.
16. Ash, B., Nalajala, V.S., Popuri, A.K., Subbaiah, T., Minakshi, M., Perspectives on nickel hydroxide electrodes suitable for rechargeable batteries: Electrolytic vs. chemical synthesis routes. *Nanomater. (Basel)*, *10*, 9, 1878, 2020, https://doi.org/10.3390/nano10091878.
17. Liu, X., Li, K., Li, X., The electrochemical performance and applications of several popular lithium ion batteries for electric vehicles: A review, 2018.
18. Ambroz, F., Macdonald, T., Nann, T., Trends in aluminium-based intercalation batteries. *Adv. Energy Mater.*, *7*, 1602093, 2017, https://doi.org/10.1002/aenm.201602093.
19. Balasubramaniam, S., Mohanty, A., Balasingam, S.K., Kim, S.J., Ramadoss, A., Comprehensive insight into the mechanism, material selection and

performance evaluation of supercapatteries. *Nano Micro Lett.*, *12*, 1, 85, 2020, https://doi.org/10.1007/s40820-020-0413-7.
20. Panic, V., Electrochemical supercapacitors: Operation, components and materials. *Chem. Industry.*, 10, 9, 1–22, 2018.
21. Czagany, M., Hompoth, S., Keshri, A.K., Pandit, N., Galambos, I., Gacsi, Z., Baumli, P., Supercapacitors: An efficient way for energy storage application. *Mater. (Basel)*, *17*, 3, 702, 2024, https://doi.org/10.3390/ma17030702.
22. Azidin, F., Hannan, M.A., Mohamed, A., An energy management of light electric vehicle. *Int. J. Smart Grid Clean Energy*, *2*, 2, 271–276, 2013, https://doi.org/10.12720/sgce.2.2.271-276.
23. Rampai, M.M., Mtshali, C.B., Seroka, N.S., Khotseng, L., Hydrogen production, storage, and transportation: Recent advances. *RSC Adv.*, *14*, 10, 6699–6718, 2024, https://doi.org/10.1039/d3ra08305e.
24. El-Eskandarany, M.S., Recent developments in the fabrication, characterization and implementation of MgH2-based solid-hydrogen materials in the Kuwait Institute for Scientific Research. *RSC Adv.*, *9*, 18, 9907–9930, 2019, https://doi.org/10.1039/c9ra00287a.
25. Mitrašinović, A.M. and Radosavljević, M., Photovoltaic materials and their path toward cleaner energy. *Global Challenges*, *7*, 2, 2200146, 2022, https://doi.org/10.1002/gch2.202200146.
26. Guerra, N., Guevara, M., Palacios Arias, C., Crupi, F., Operation and physics of photovoltaic solar cells: An overview. *I+D Tecnológico*, *14*, 2, 84–95, 2018, https://doi.org/10.33412/idt.v14.2.2077.
27. Zheng, J.-C., Recent advances on thermoelectric materials. *Frontiers of Physics in China*, *3*, Article 0028, 2011, https://doi.org/10.1007/s11467-008-0028-9.
28. Parvathy, T. and Pradyumnan, P.P., Impact of mobility and effective mass on the thermoelectric performance of Ni doped Cu2Se. *J. Alloys Compd.*, 970, 172615, 2024, https://doi.org/10.1016/j.jallcom.2023.172615.
29. Chen, T.-W., Anushya, G., Chen, S.-M., Kalimuthu, P., Mariyappan, V., Gajendran, P., Ramachandran, R., Recent advances in nanoscale based electrocatalysts for metal-air battery, fuel cell and water-splitting applications: An overview. *Materials*, *15*, 2, 458, 2022, https://doi.org/10.3390/ma15020458.
30. Wang, X., Wang, J., Sun, Y., Li, K., Shang, T., Wan, Y., Recent advances and perspectives of CeO2-based catalysts: Electronic properties and applications for energy storage and conversion. *Front. Chem.*, *10*, 1089708, 2022, https://doi.org/10.3389/fchem.2022.1089708.
31. Zhao, Y., Yang, N., Yu, R., Zhang, Y., Zhang, J., Li, Y., Wang, D., Unique structural advances of graphdiyne for energy applications. *EnergyChem*, *2*, 5, 100041, 2020, https://doi.org/10.1016/j.enchem.2020.100041.
32. Sun, N. (Ed.), *Advanced materials for energy storage and conversion*, MDPI, Basel, Switzerland, 2024, https://doi.org/10.3390/books978-3-0365-9996-0.
33. Nugroho, A.A., Liu, Y., Umemura, K. (Eds.), *Composite materials for energy conversion and storage applications*, Trans Tech Publications, Stafa-Zurich, Switzerland, 2023, https://doi.org/10.4028/b-52rrMO.

34. Zhao, S., Wang, H., Ruediger, A., Gu, F., Yan, D., Lu, J., Editorial: Micro/nano materials for energy storage and conversion. *Front. Chem.*, *11*, 1150885, 2023, https://doi.org/10.3389/fchem.2023.1150885.
35. Hone, F., Tegegne, N., Andoshe, D., Advanced materials for energy storage devices, in: *Advanced Materials for Energy Storage Devices*, pp. 71–107, CRC Press, Boca Raton, Florida, USA, 2021, https://doi.org/10.1201/9781003145585-5.

10

Innovations in the Synthesis of Nanomaterials: Cutting-Edge Techniques Along with the Diverse Implementations of These Nanomaterials in Nanotechnology Methods

K.K.A. Rinsila[1,2*], Anu Ruby Benny[2] and Lina Anil[2]

[1]Faculty of Ocean Science and Technology (FOST), Kerala University of Fisheries and Ocean Studies (KUFOS), Panangad, Kerala, India
[2]Department of Biosciences, MES College Marampally, Ernakulam, Kerala, India

Abstract

Nanoparticle synthesis stands at the forefront of modern material science, offering unparalleled potential in various fields, ranging from electronics to medicine. An extensive exploration of the latest nanoparticle synthesis trends, with a focus on both bottom-up as well as top-down methods, is provided in this chapter. In the former method, chemical and biological approaches take center stage, harnessing the principles of self-assembly and biomineralization to fabricate nanoparticles by having accurate control over composition, size, and shape. Chemical methods enable the production of nanoparticles with tailored properties through controlled nucleation and growth processes. Biological methods, mediated by plant extracts and microorganisms offer sustainable and eco-friendly routes to nanoparticle synthesis, emphasizing the importance of biocompatibility and biodegradability. Conversely, the latter method employs physical as well as chemical techniques to fabricate nanoparticles by downsizing bulk materials. While physical methods enable the production of nanoparticles through mechanical or thermal energy, chemical methods provide precise control over nanoparticle dimensions through patterned growth and selective deposition. By elucidating the latest advancements

**Corresponding author*: rinsilafazil104@gmail.com

Kolla Bhanu Prakash, S. V. Ranganayakulu and K. S. Jagannatha Rao (eds.) New Frontiers in Materials Science: Interdisciplinary Approaches to Innovation and Technologies, (187–214) © 2025 Scrivener Publishing LLC

in nanoparticle synthesis across bottom-up and top-down methodologies, this chapter offers valuable insights into the diverse strategies employed to tailor nanoparticle properties for a myriad of applications, from catalysis and sensing to drug delivery and nanoelectronics.

***Keywords*:** Nanoparticle synthesis, bottom-up method, chemical method, biological method top-down methods, physical method

10.1 Introduction

Tiny particles with dimensions ranging from 1–100 nanometers are called nanoparticles. In recent years, nanomaterials have gained more interest owing to their large number of applications [34]. Their higher aspect ratio enhances a myriad of characteristics, including optical, chemical, mechanical, electrical, thermal, and antibacterial capabilities, which were different from those of their macroscopic nature. The concept of nanobiotechnology has emerged as a potential alternative for biological applications. Nanoparticles are often applied in environmental, biological, and medical uses due to their biological absorption, tumor targeting, bioavailability, bioactivity, effective drug delivery, antimicrobial, anti-inflammatory action, and biocompatibility [26]. There are several techniques for creating nanoparticles, including biological, physical, and chemical processes. The type of nanoparticles needed, their intended size and shape, and the resources available are the factors that determine the choice of the

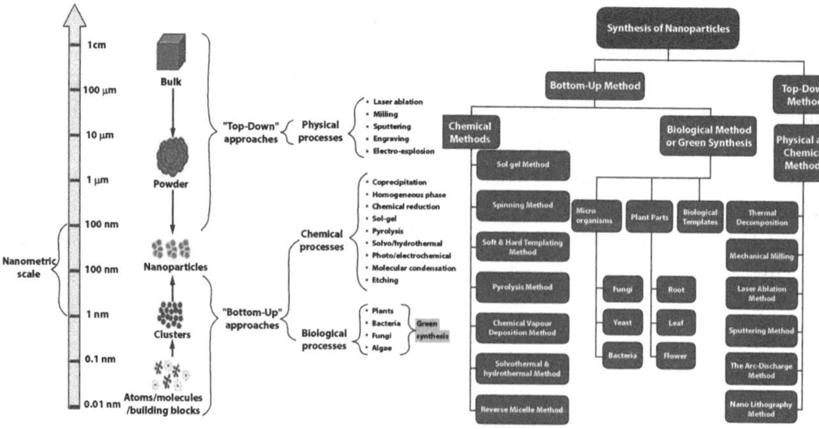

Figure 10.1 Different approaches to nanomaterial synthesis and nanometric scale.

particular method. The main strategies for the synthesis of nanoparticles include Bottom-Up as well as Top-Down techniques (Figure 10.1).

10.2 Bottom-Up Method

The method involves putting together nanoparticles (NPs) from fundamental units of smaller particles, molecules, and atoms, which results in more intricate assemblies. This approach reveals three different alternatives:

1. Chemical synthesis: This approach relies on the reaction of raw materials to produce molecules or particles.
2. Self-assembly: In this technique, molecules or atoms organize themselves through chemical and/or physical interactions.
3. Positional assembly: In this process, aggregates, molecules, and atoms are intentionally manipulated and arranged individually.

Various methods are commonly employed, such as wet-chemical processes, solvothermal, microemulsion, chemical vapor deposition (CVD), co-precipitation, radiation-induced, pyrolysis, hydrothermal, electrodeposition, and solid-gel chemistry techniques. Though there are disadvantages of ecotoxicity and expensiveness, this method is favored more than the top-down method as it doesn't require specialized equipment and allows for faster production of nanoparticles. Green synthesis is becoming increasingly important for the production of NPs using the bottom-up method.

10.3 Chemical Method

10.3.1 Sol–Gel Method

Using a wet technique, this method is used to develop diverse nanomaterials, especially metal oxide nanoparticles. The most frequently utilized precursors are chlorides and metal oxides. Figure 10.2 shows an overview of the different kinds of processes that can be done with the sol-gel method. Metal alkoxides, which can dissolve in alcohol or water and form a gel when stirred and heated through alcoholysis or hydrolysis, are frequently used as molecular precursors. The resulting gel is moist or wet, and drying it utilizing suitable techniques determined by its preferred characteristics as well as its intended use is crucial. For instance, the drying procedure includes burning off the alcohol when the gel is an alcoholic solution.

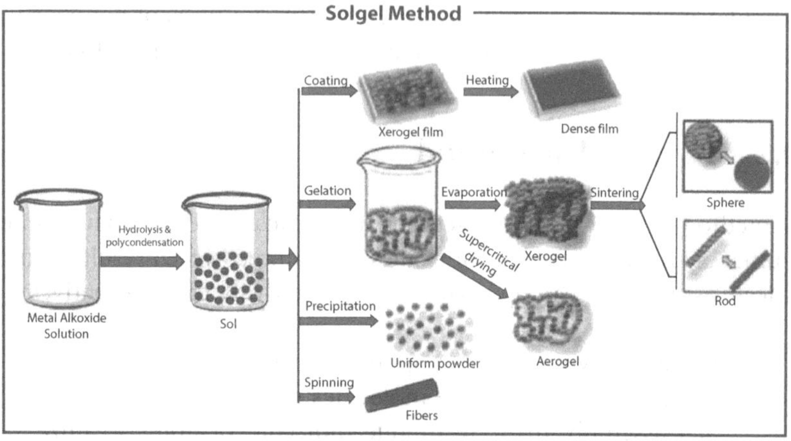

Figure 10.2 An overview of the types of processes that can be done with the sol–gel method and the product of each process [13].

Once dried, the gels are pulverized and then subjected to calcination. The sol–gel process transforms a uniform sol, created from precursors, resulting in a gel. The residual material is dried after subsequently removing the solvent from the gel formation. The drying technique greatly impacts the characteristics of the gel that is dried. In simpler terms, the drying technique depends on the intended use of the gel. Diverse industries use these dried gels, including the production of specialty clothing, surface coating, and building insulation. It's crucial to note that special mills can be used to grind the gel into nanoparticles [11].

10.3.1.1 Applications and Advantages of the Sol–Gel Method

1. Nanoparticle synthesis involves traditional and industrial methods to create nanoparticles with diverse chemical compositions.
2. The affordability of nanoparticle synthesis, coupled with the ability to precisely control chemical constituents despite low-temperature synthesis, makes it a favorable approach.
3. Nanoparticles synthesized through this method find applications as molding materials in ceramic manufacturing and as intermediaries in various applications involving delicate films of metal oxides.
4. This process yields materials utilized across a spectrum of applications, including electrical, energy, optical, bio-

sensors, surface engineering, separation technologies, and pharmaceuticals.

5. This method allows for the creation of amorphous materials and their application in thin layers.
6. Nanoparticles produced exhibit varied physical properties like reduced CTE (coefficient of thermal expansion), high optical transparency, and low UV absorption.
7. The production process also yields porous materials suitable for enrichment with organic and polymeric substances [11].

10.3.2 Spinning Method

In the approach of producing NPs via chemical means, this spinning method entails the rapid blending of precursor solutions or reagents by harnessing centrifugal force. This technique commonly involves the use of a spinning disk or rotating container to induce high shear rates and facilitate thorough mixing, thereby fostering the nucleation as well as the development of NPs.

Precursor Solutions Preparation: This process commences with the preparation of precursor solutions containing the requisite chemicals or reactants for nanoparticle formation. These solutions typically comprise metal salts, reducing agents, stabilizing agents, and solvents.

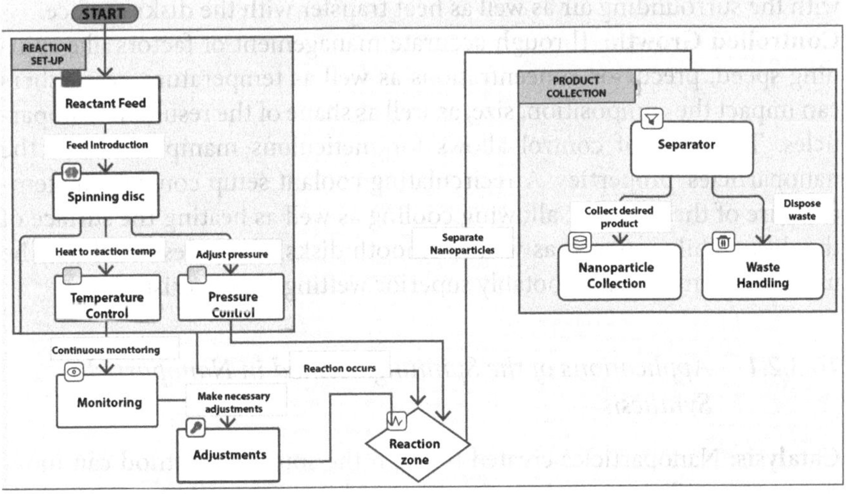

Figure 10.3 Schematic representation of a spinning disk reactor.

Loading onto the Spinning Platform: The precursor solutions are placed onto the rotating platform, which may take the form of a spinning disk, rotating container, or similar configuration. In spinning disk processing (see Figure 10.3), a stainless-steel disk that rotates rapidly (at 5–3000 rpm) is utilized, with reagents being dispensed onto it through several jets. Subsequently, the platform is set into motion, typically at high speeds.
Centrifugal Force: As the rotational platform undergoes spinning, centrifugal force is exerted on the precursor solutions. Centrifugal forces cause the reagents to move across the disk's surface, creating profoundly efficient turbulent micromixing. The force causes the solutions to radially disperse, forming a thin film or layer on the spinning platform's surface. Continuous gear pumps are used to deliver low-viscosity solvents, with the ideal throughput for the SDP ranging from 0.5–3 ml/s. In the reactor, the residence times are usually under 1 second. The formation of rivulets, rather than the required fluid film, may occur with reduced disk speeds or feed rates. On the other hand, higher feed rates would necessitate an impractically extensive spin-up zone.
Enhanced Mixing: The rapid spinning movement induces vigorous mixing of the initial solutions, leading to the uniform dispersion of reactants and acceleration of chemical reactions. This process promotes the formation and expansion of nanoparticles within the solution. At the center of the spinning disk, the reactant solutions are added. Here, until a reverse hydrostatic jump occurs, a viscous drag drives them outward, which causes the fluid coating to disperse throughout the disk within a thickness of 10–200 μm. This delicate, quickly shifting coat facilitates adequate mass transfer with the surrounding air as well as heat transfer with the disk's surface.
Controlled Growth: Through accurate management of factors like spinning speed, precursor concentrations as well as temperature, researchers can impact the composition, size, as well as shape of the resultant nanoparticles. This level of control allows for meticulous manipulation of the nanoparticles' properties. A recirculating coolant setup controls the temperature of the reaction, allowing cooling as well as heating the surface of the disk. While grooved as well as smooth disks of stainless steel may be used, the former exhibit notably superior wetting characteristics.

10.3.2.1 Applications of the Spinning Method in Nanoparticle Synthesis

Catalysis: Nanoparticles created through the spinning method can function as very efficient catalysts for diverse chemical reactions due to their exceptional surface attributes as well as elevated SA: V (surface area to volume) ratio.

Biomedical Imaging: Nanoparticles with tailored optical properties, like plasmonic nanoparticles or quantum dots, synthesized via the spinning method, are utilized by biomedical imaging techniques such as fluorescence imaging as well as SERS (Surface-Enhanced Raman Scattering) as contrast agents.

Drug Delivery: NPs synthesized utilizing the spinning method can undergo functionalization with targeting ligands and be loaded with therapeutic agents, rendering them ideal for targeted drug delivery applications.

Sensor Technology: Nanoparticles produced through the spinning method are employed in the generation of sensors for detecting different analytes, including gases, biomolecules, and environmental pollutants, owing to their high sensitivity and selectivity.

Energy Storage and Conversion: Nanoparticles synthesized via the spinning method find applications in appliances for energy storage (e.g., supercapacitors and batteries) as well as technologies for energy conversion (e.g., fuel cells and solar cells) due to their unique electronic and electrochemical properties.

10.3.3 Template Method

In the methodology known as the template method, the organic molecular structure of the synthetic material serves as a template, manipulating its morphological properties, framework, and dimensions through forces like hydrogen bonding and Van der Waals. The template method essentially involves a molding or shaping process [16]. Broadly, the production of the template technique comprises three key stages [42]: a) preparation of the template; b) target materials synthesis based on the template; and c) template removal.

Recently, there have been several template agent classes; each category is indicated in Figure 10.4. A single type of category may be classified into soft templates (like surfactants and biomolecules) and hard templates (like silica and porous silica) based on the spatial domain-limiting abilities as well as the characteristics of the template [18]. The soft template approach is a basic conventional technique for generating nanostructured materials [19]. The most significant interactions between the precursors and the soft templates happen via van der Waals, electrostatic forces, and hydrogen bonding [5]. Flexible organic molecules, block copolymers, non-ionic, cationic, and anionic surfactants are the most widely used soft templates [20]. 3D organized mesoporous systems are created using soft templates made of precisely structured liquid crystalline particles. Several parameters like concentrations of precursor and surfactant, surfactant framework, environmental conditions, and surfactant-to-precursor ratio may impact the mesoporous material systems formed out of 3D organized micellar structures [19]. The soft template approach may yield several nanostructured materials, such as mesoporous N-doped graphene [22], porous aluminas [6], single-crystal nanorods [21], and polymeric MCSs (mesoporous carbonaceous nanospheres) [20]. Efficient solid materials serve as templates in the case of the hard template process (or nano-casting), and precursor molecules are loaded into the pores of this solid template to produce nanomaterials in favor of specific purposes [5]. To create structured mesoporous materials, selecting a rigid template is essential. The mesoporous structure of the aforementioned hard templates must be preserved throughout the conversion process of the precursor, and they must be effortlessly extractable without altering the final nanostructure. Wood shells, colloidal crystals, particles, carbon nanotubes, silica, and carbon black are the most commonly utilized hard templates [23].

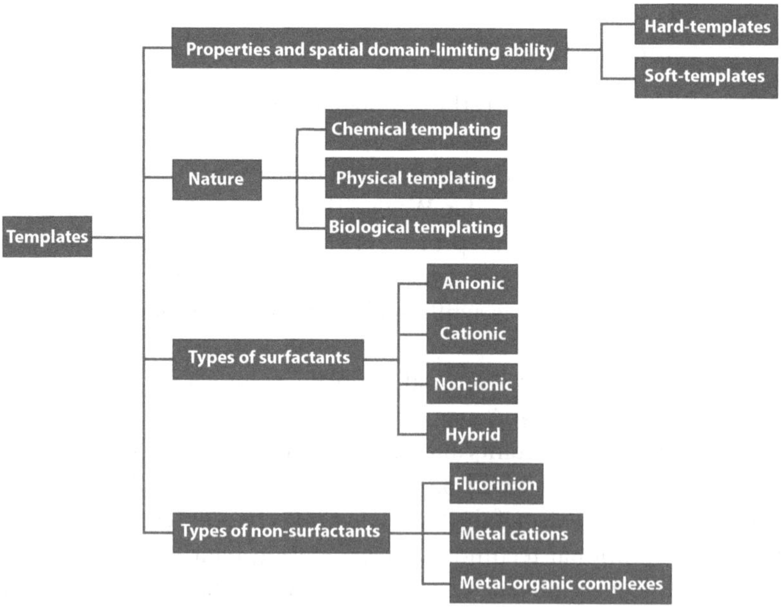

Figure 10.4 Classification of templates [15]. Copyright © 2022 Li, Wang, Wei, Yan, and Feng.

The categorization of templates encompasses physical, chemical, and biological classifications based on their inherent characteristics [19, 20]. Furthermore, certain researchers subdivide templates into non-surfactants and surfactants. The former category comprises metal-organic complexes, organic molecules, metal cations, and similar entities, while the latter includes hybrid, non-ionic, cationic, and anionic template agents (Figure 10.4) [15].

10.3.3.1 *Applications of Template Methods*

1. Owing to the production of varied morphological substances, typically mild experimental conditions, as well as its simple implementation, the soft template method is deemed useful [19].
2. Hard templates provide superior domain limits and stability compared to soft templates, allowing for stricter morphology and material size regulation. As per the necessities of the desired outcome to obtain exact control of distinct morphology, pore diameters, particular surface areas, and other properties, alternative template agents can be chosen [15].

10.3.4 Laser Pyrolysis

This method represents a highly efficient gas-phase approach to produce a diverse range of NPs at a laboratory level. The fundamental concept of this technique involves the decomposition of gaseous or liquid reactants through the application of a CO2 laser which is of elevated power, followed by rapid cooling. Recent research indicates that this process can result in the production of nanoparticles encompassing composites, metals, oxides, nitrides, and carbides. In a usual laser pyrolysis procedure (see Figure 10.5), gaseous-phase precursors are delivered into a chamber with the aid of a carrier gas like argon, where they subsequently interact with the laser beam. The utilization of a high-power laser beam, for instance, at 2400 W, induces localized temperature increases, leading to nanoparticle nucleation and subsequent growth. The resulting nanoparticles are gathered using a catcher fitted with a filter. Although laser pyrolysis is categorized as a gaseous-phase process, it can accommodate the use of liquid reactants where necessary. These liquid reactants can be vaporized or transformed into microscale droplets, for example, through an ultrasonication technique, before being introduced into the reaction chamber. Given that this technology heavily relies on the thermal energy generated by high-power lasers, continuous-mode lasers such as CO2 are commonly employed [24].

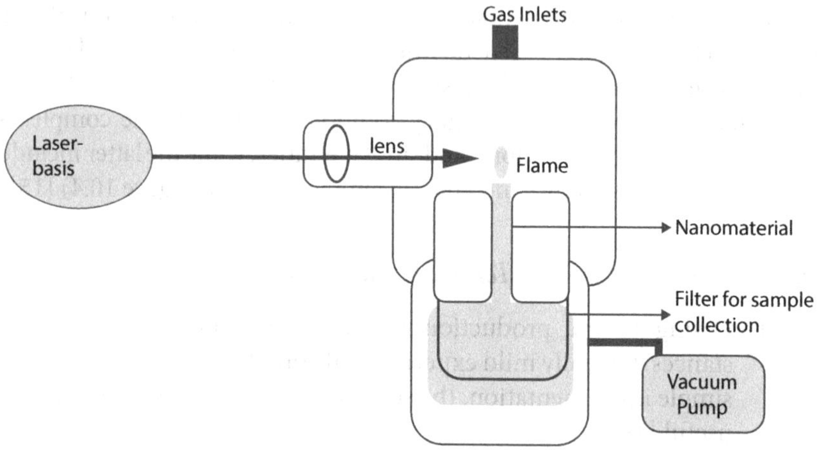

Figure 10.5 Laser pyrolysis.

10.3.4.1 *Applications of Laser Pyrolysis*

The laser pyrolysis technique is widely utilized for nanoparticle production and coating. This technology has been implemented across various scales and methodologies, resulting in a diverse array of outputs. Notably, recent advancements have led to the exploration of new research avenues, particularly in the realms of biomedical applications, additive manufacturing, and waste disposal [25].

10.3.5 Chemical Vapor Deposition Method

The chemical vapor deposition (CVD) method is a hybrid method using chemicals in the vapor phase and reactions in a controlled environment to produce nanoparticles with high purity and specific characteristics. The gaseous reactant is transferred to the substrate kept at an elevated temperature. On contact, the formed products undergo a chemical reaction at a specific site, nucleate, and elongate to form the desired material film. The accumulated by-products can be removed in the gaseous form from the substrate [38]. The reaction chamber should be replenished with vapors of required gases. CVD proceeds through profuse processes such as the reduction of gas, chemical reactions between different source gases, oxidation, or some disproportionate reaction. Usually, temperatures extending from 300 to 1200°C are used at the substrate. The substrates are heated by either a cold wall or a hot wall.

In the hot wall approach, the deposition takes place on the reactor wall; moreover, a reaction occurs in the gas phase, which is avoided in the cold wall method. Furthermore, at a gas pressure in the range of 0.1 torr to 1.0 torr, coupling of plasma with a chemical reaction is feasible in a cold wall setup. The growth rate and film quality are influenced by gaseous pressure and the temperature of the substrate. Nanocrystalline films or single crystalline films are possible. Based on the difference in source gas pressure, geometrical layout, and temperature used, several variants of CVD, such as Metallo organic CVD (MOCVD), atomic layer epitaxy (ALE), vapor phase epitaxy (VPE), plasma enhanced CVD (PECVD), etc., have been developed. Overall, CVD offers advantages such as relatively simple instrumentation, ease of processing, the possibility of depositing different types of materials, economic viability, high product purity, and scalability [37].

10.3.6 Hydrothermal Method

The hydrothermal/solvothermal method represents a wet chemical method employed for crystallizing the materials to a nanostructure. The process is carried out in a sealed container, called a Hydrothermal Autoclave Reactor, under high vapor pressure and temperature (ranging from 100°C–250°C) in an aqueous solution. The synthesis of single crystals depends on the solubility of minerals in hot water under high pressure. The nutrient is supplied along with water in the autoclave. Opposite ends of the growth chamber maintain a temperature gradient. The nutrient solute dissolves at the hotter end, while at the cooler end, seed crystal deposition occurs, and subsequently, the desired crystal grows [39].

The methods used in hydrothermal synthesis and crystal growth include:

1. Temperature-difference method: A temperature gradient is created in the autoclave, thereby super-saturation is achieved in the crystal growth zone.
2. Temperature-reduction method: There is no temperature differential between the growth and dissolution zones during crystallization. Super-saturation is accomplished by gradually lowering the solution's temperature inside the apparatus.
3. Metastable-phase technique: This technique relies on the difference in solubility between the initial material and the phase to be produced. The nutritional material is more soluble than the stable phase, and the metastable phase dissolves, causing the stable phase to crystallize.

10.3.6.1 Benefits of Hydrothermal Synthesis

1. Ability to induce crystalline phases at any melting point safely.
2. Fabricate substances with vapor pressures close to their melting points.
3. Generates high-quality, bigger crystals and nanoparticles with control over their composition and content.
4. Relatively simple and green process, making it environmentally friendly.
5. Disadvantages include the requirement of an expensive autoclave and the limitation in crystal growth observation inside the tube [35].

10.3.6.2 Uses

- Synthesis of a large number of compounds, synthetic quartz, and other single crystals with high commercial value is possible under hydrothermal conditions.
- Innovative substances with defined physical characteristics and the advancement of methodical physicochemical investigation of complex multicomponent systems at high pressure and temperature [36].

10.3.7 Reverse Micelle Method

One of the exciting methods for producing nanoparticles with precise size and shape is the reverse micellar technique. Reverse micelles (RMs) are formed by a minimum of three components: two immiscible phases (organic and aqueous) and a surfactant with amphiphilic properties [33]. These reverse micelles consist of tiny water droplets dispersed uniformly within a nonpolar organic solvent, surrounded by a well-defined, dynamic surfactant layer. RMs form when surfactant molecules spontaneously self-assemble in nonpolar organic solvents, creating dry micelles. They can exist in various ternary systems. When water or another polar solvent is introduced, the micelles absorb the solvent into their hydrophilic core, causing the micelles to swell. Due to certain similarities, this system is often referred to as water-in-oil microemulsions, though the boundary between the two is somewhat flexible. A key characteristic of RMs is their ability to serve as reaction media, dissolving substances of varying polarities, including precursors for the controlled synthesis of metal nanoparticles [28].

10.4 Green Synthesis or Biological Method

The Green synthesis method shown in Figure 10.6 utilize biological agents such as viruses, bacteria, fungi, algae, and plants for the synthesis of nanoparticles. Nonpathogenic organisms are used for the synthesis because they never interfere with the application of synthesized nanoparticles [30]. Compared to conventional synthesis methods, this approach has a number of benefits, such as lower costs, scalability, and less hazardous waste. Green synthesis enables the creation of nanoparticles with distinct sizes, shapes, and surface characteristics that have been altered for particular purposes. Enzymes, proteins, polyphenols, flavonoids, and terpenoids are among the biologically active substances found in the biological sources used for the environmentally friendly synthesis of nanoparticles. These substances can function as catalyzing, reducing, stabilizing, or capping agents for one-step synthesis [27].

10.4.1 From Roots

The fabrication of metal nanoparticles through green synthetic pathways using plant extracts has gained increased attention due to reliable, sustainable, eco-friendly protocols and applications. The ability of plant extracts to reduce metal ions has been known since the early 1900s, although the nature of the reducing agents involved was not well understood. Through green synthesis techniques, Biomolecules present in plant extracts can be used to reduce metal ions to nanoparticles in a single step has attracted considerable attention. This biogenic reduction of metal ions to base metal is quite rapid and can be readily conducted at room temperature and

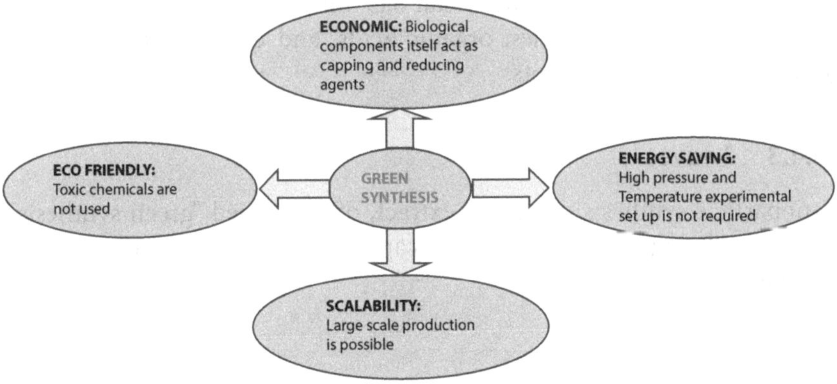

Figure 10.6 Perks of green synthesis.

pressure. Plants are a good and easily available source of bioactive plant secondary metabolites such as polysaccharides, proteins, polyphenols, flavonoids, terpenoids, tannins, alkaloids, amines, ketones, and aldehydes, which act as reducing, stabilizing, and capping agents in the conversion of metal ions to metal nanoparticles, leading to the production of desirable nanoparticles with predefined characteristics [40].

The typical method for green synthesis of metal/metal ions NPs is to use reducing agents from plant extracts. The extracts are usually obtained by soaking ground plants in solvents like water or ethanol under suitable environmental conditions (different green substances have different optimum conditions). The extracts are then mixed with a solution containing the desired metal ions. The synthesis of gold (Au) NPs turns the solution red, and the synthesis of silver (Ag NPs) is indicated by turning the solution brownish. Although the size and shape of nanoparticles synthesized by different extracts are highly variable, the most common shapes of green-synthesized NPs are spheres, and some are triangles and hexagons. Among the several factors affecting the green synthesis of plant extracts, pH and temperature are more prominent. Different parts of the same plant can have different effects [41].

The difference in particle size, or controlling particle size during production, and the lack of technique to determine the properties are major challenges in green technology. Innovations are required to improve the conversion rate and yield of nanoparticles prepared from green materials.

10.4.2 Flowers

Plant parts are used in research to reduce and stabilize nanoparticles. Metal or metal oxide nanoparticles are synthesized from various plant parts, including flowers. Flowers contain proteins, coenzymes, and carbohydrates, which reduce the metal salt into nanoparticles. Secondary metabolites such as flavones, organic acids, and quinones in plants are good natural reducing agents.

10.4.3 Leaves

Nanoparticle synthesis using leaf extract, often termed "green synthesis," involves utilizing plant extractions as capping and reducing agents to yield nanoparticles.

Green Synthesis Mechanisms using Leaf Extract (Figure 10.7):

- Selection of Plant Material: Various plants are chosen based on their phytochemical composition, which acts as reducing and stabilizing agents in the nanoparticle synthesis. Common choices are neem, tulsi, green tea, and aloe vera.
- Preparation of Leaf Extract: Leaves are gathered, cleansed, and prepared to acquire an aqueous derivative. This extract contains bioactive blends such as phenolics, terpenoids, and flavonoids, which serve as reducing agents.
- Nanoparticle Formation: The leaf extract is mixed with metal salt precursors (e.g., silver nitrate, gold chloride) under controlled conditions (temperature, pH). The metal ion reduction occurs due to the phytochemical presence, leading to nanoparticle formation.
- Characterization: To define their shape, size, crystallinity, as well as stability, procedures such as UV-Vis spectroscopy, X-ray diffraction (XRD), transmission electron microscopy (TEM), and dynamic light scattering (DLS) are utilized to characterize the produced NPs.

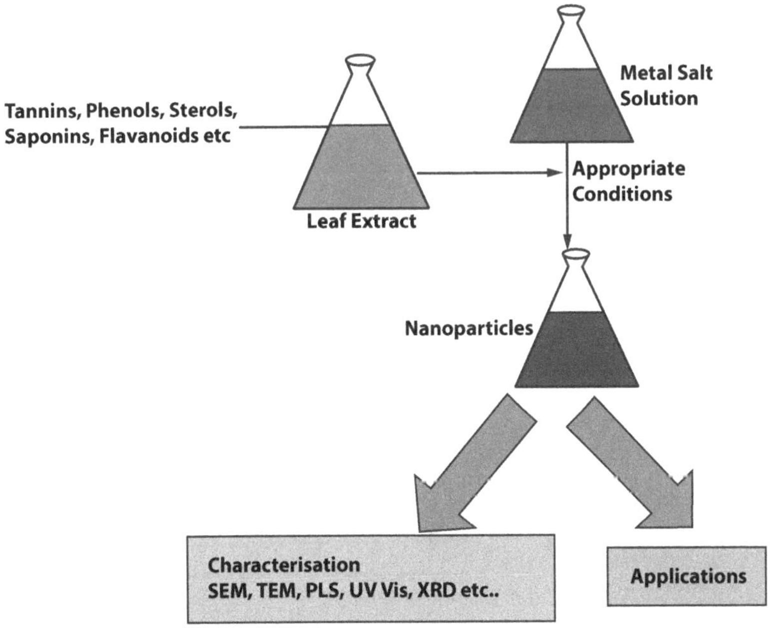

Figure 10.7 Schematic representation of NP green synthesis utilizing extract from the leaf.

Plants used (Figure 10.8): The synthesis of silver and gold NPs utilizing extracts of Geranium, sun-dried *Cinnamomum camphora*, and *Azadiracta indica* [12] leaves are demonstrated. Additionally, the cost-effective reduction of gold and silver ions in solution, when exposed to extracts of a plant leaf, leads to the formation of bimetallic silver–gold shell NPs. Nanoparticle synthesis is furthermore carried out using *Murraya koenigii* leaf extract.

Applications:

1. Biomedical: Nanoparticles synthesized using leaf extract have shown promise in imaging, therapeutics, as well as drug delivery due to their biocompatibility and controlled release characteristics.
2. Environmental Remediation: These NPs are utilized for the degradation of pollutants like heavy metals, pesticides, and dyes, due to their catalytic and adsorption capabilities.
3. Food Packaging: Nanocomposites derived from leaf-extract-synthesized NPs are explored for antimicrobial effects, improving the longevity of food products.

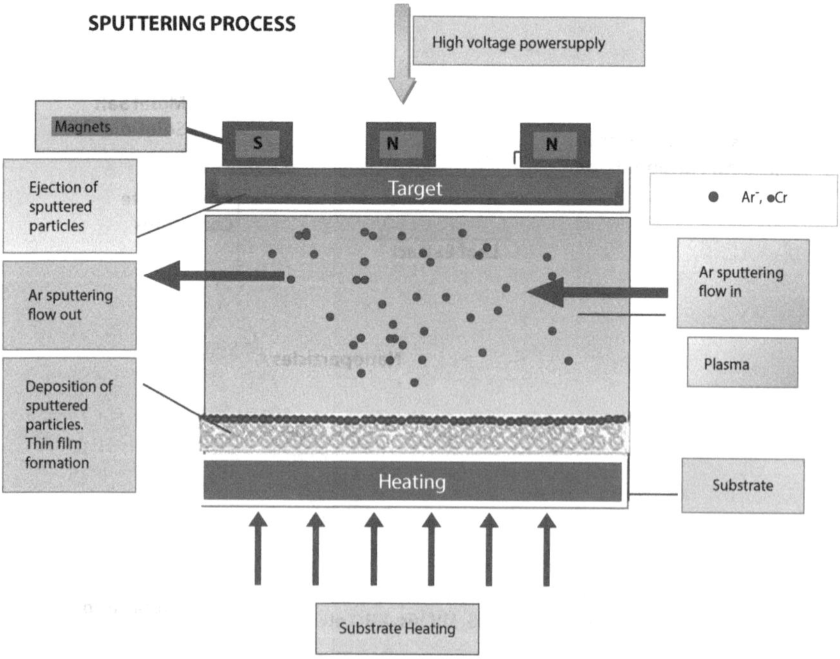

Figure 10.8 Sputtering method.

4. Catalysis: Leaf-extract-mediated nanoparticles function as efficient catalysts in diverse organic reactions, offering a sustainable alternative to conventional catalysts.
5. Textile Industry: Nanoparticles synthesized using plant extracts are incorporated into textiles to impart antimicrobial and UV-blocking properties.

10.4.4 Bacteria

Bacterial nanoparticle synthesis involves the utilization of microorganisms, primarily bacteria, to convert metal ions into nanoparticles through reduction or bioaccumulation processes. This method delivers several advantages, including the ability to synthesize NPs of diverse sizes and shapes, cost-effectiveness, as well as eco-friendliness.

Mechanisms of Bacterial Synthesis (Figure 10.9):
Enzymatic Reduction: Certain bacterial enzymes can reduce metal ions into their nanoparticle forms. For example, enzymes like nitrate reductase, nitrite reductase, and hydrogenases can reduce metal ions like silver, gold, and platinum into nanoparticles.

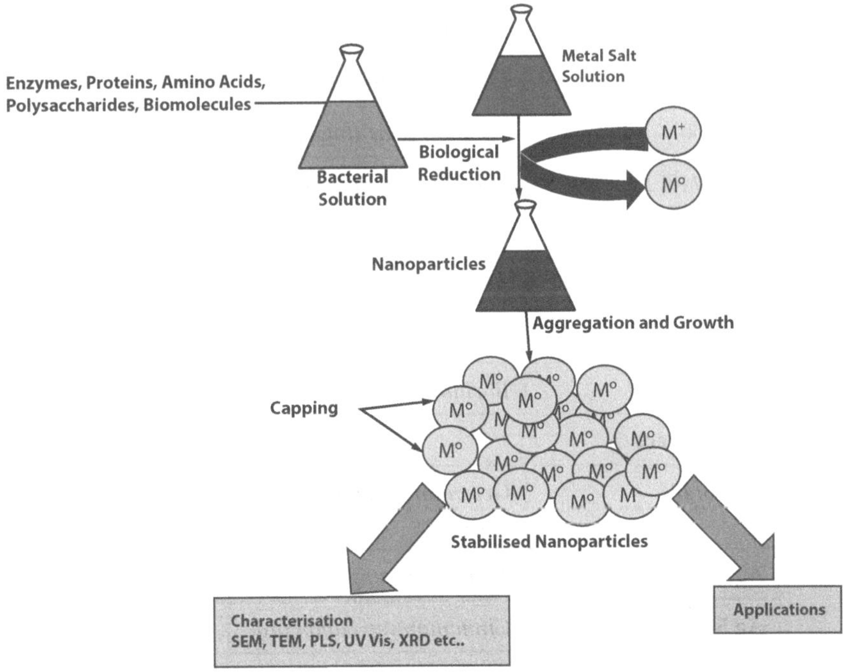

Figure 10.9 Schematic representation of biological synthesis of NPs using bacteria.

Genetic Engineering: Genetic manipulation of bacteria can be employed to enhance their nanoparticle synthesis capabilities. This can involve introducing or modifying genes involved in metal ion reduction, biomolecule production, or nanoparticle assembly to improve synthesis efficiency and yield.

Bacterial Strains Used (Table 10.1): Various bacterial strains have been employed for the NP synthesis, including *Bacillus*, *Escherichia*, *Pseudomonas*, and others. Each strain offers unique capabilities and characteristics for nanoparticle synthesis.

Factors Influencing Synthesis:
pH: The pH of the reaction environment can significantly influence nanoparticle synthesis by affecting bacterial metabolism and enzyme activity.

Table 10.1 Sources and references of bacterial and green synthesis of nanoparticles.

Type and size of NP	Source	References
Bacteria		
Ag 150/180	*Staphylococcus aureus*	[50]
Ag 5/15	*Corynebacterium*	[49]
Ag>200	*Pseudomonas stutzeri*	[48]
Au 15/40, Ag	*Lactobacillus strains*	[47]
Cds	*Escherichia coli*	[46]
Ag 30/50	*Escherichia coli*	[45]
Ag 10/20	*Bacillus thuringiensis*	[44]
Ag 5	*Bacillus cereus*	[43]
Leaf extract		
Ag 16/40	*Geranium* plant leaf extract	[12]
Ag	*Murraya koenigii*	[17]
Ag 50	*Cinnamomum camphora*	[14]
Ag, Au 50/100	*Azadirachta indica*	[12]

Temperature: Optimal temperature conditions are crucial for bacterial growth and enzymatic activity, thereby impacting nanoparticle synthesis. Metal Ion Concentration: The size distribution and formation rate of NPs are determined by the reaction mixture's metal ion concentration.

Characterization Techniques: Diverse analytical methods are used to characterize bacterial-synthesized NPs, such as dynamic light scattering (DLS), UV-Vis spectroscopy, Fourier-transform infrared spectroscopy (FTIR), X-ray diffraction (XRD), transmission electron microscopy (TEM), and scanning electron microscopy (SEM).

Applications

1. Biomedical Applications: Bacterial-synthesized NPs have demonstrated potential in biomedical uses like imaging, drug delivery, diagnostics, and therapy due to their biocompatibility and tunable properties.
2. Electronics: Nanoparticles synthesized using bacterial methods may be employed in nanoelectronics, electronic devices, and sensors.
3. Catalysis: Bacterial-synthesized nanoparticles exhibit catalytic properties and can be employed in diverse chemical reactions as catalysts.
4. Environmental Remediation: These NPs may be utilized for environmental remediation objectives, including the elimination of heavy metals and contaminants from soil and water.

10.5 Top-Down Method

The 'top-down' and 'bottom-up' techniques are the two procedures to manufacture nanoparticles. In the top-down technique, nanoparticles are made by dissolving bulk materials into small particles using lithography, sputtering, mechanical and chemical etching, thermal evaporation, pulsed laser ablation, and photo reduction methods [31].

Advantages of a Top-Down Approach:

1. Cost and controlled shape and size of the product.
2. Consistency in terms of structural attributes such as particle shape, size, and geometry.

3. It provides a monodisperse population with improved yields, enabling the scale-up of manufacturing.
4. It improves encapsulation efficiency and drug stability.
5. Useful for making macroscopic connections by producing structures with long-range order.
6. The major disadvantage is the incomplete surface structures.

10.6 Physical and Chemical Methods

10.6.1 Thermal Decomposition

In this method, heat energy is utilized for the breaking of chemical bonds. Nanoparticles are synthesized by the decomposition of metal particles at specific temperatures and produce secondary products. The optimum temperature maintained is called the decomposition temperature. Hydrothermal and solvothermal are the two methods that work in a closed format [9]. This method has an extensive application in engineering nanomaterials and producing various patterns of materials including nanowires, nanorods, nanosheets, and nanospheres [10].

10.6.2 Mechanical Milling

Among all top-down approaches, mechanical milling has been extensively used for the synthesis of various nanomaterials. Mechanical milling aims to reduce particle size and blend particles into new phases. The ball milling technique is a versatile mechanochemical technique to synthesize nanomaterials, including metallic, multi-metallic, alloyed, and ceramic nanocomposites. The central occurrence is the ball-powder-ball collision. Metallic particles and nanomaterials are mixed and ground in a sample jar containing a large number of balls. A progressive, repeated collision between the particles, balls, and vessel walls leads to particle fracturing and cold welding. The combined effect of compression, shear, and frictional forces breaks down the materials into nano-powder. The kinetics of mechanical milling or alloying depend on the energy transferred to the powder from the balls during milling. The energy transfer is governed by many parameters, such as the type of mill, the powder supplied to drive the milling chamber, milling speed, size and size distribution of the balls, dry or wet milling, temperature of milling, and duration of milling. The temperature during milling can depend on the kinetic energy of the ball and the material characteristics of the powder and milling media. The synthesized nanomaterials can be modified by combining the mechanical milling process with post-annealing or activation processes [32].

10.6.2.1 Type of Ball Mills

1. Based on Mechanical Energy: Based on the mechanical energy induced onto the sample inside the chamber, mechano-chemical milling can be categorized as low energy (LEBM) and high energy ball mills (HEBM).
2. Based on Operation Mode: Depending on their operations, two modes are categorized: direct and indirect milling. In the former, kinetic energy (KE) is directly imparted onto the powder sample through the shafts, while in the latter, KE is imparted into the body of the mill and then transmitted to the balls and samples charged through friction.
3. Based on Wettability: Based on wettability, ball mills are classified into dry and wet mills. In wet ball mills, a surface-active media is introduced to inhibit the formation of aggregates, while dry mills are free of solvents, and the attrition of particles occurs through powder-to-powder friction [29].

The advantage of mechanochemical alloying is the homogenous distribution of compounded particles into the matrix, which prevents agglomeration. Contamination from the milling media or atmosphere is its major limitation.

10.6.3 Laser Ablation Method

As the name implies, laser ablation imparts the synthesis of nanoparticles with the help of laser beams. As a result of high-energy laser irradiation, the precursor materials vaporize. As the process is devoid of stabilizing agents and chemicals, the method can be included as a green technique [4]. It has a wide range of applications for the synthesis of carbon nanomaterials, metal nanoparticles, oxide composites, and ceramics [7]. Colloidal nanoparticles can be obtained by pulse laser ablation. The fluence of pulse laser, wavelength, and addition of laser salt efficiently contribute to the working of laser ablation. The sample is kept in a vacuum chamber.

10.6.4 Sputtering Method

Sputtering is considered to be an efficient synthesis process for the production of nano-thin films [1]. The process of sputtering consists of the bombardment of gaseous ion particles and the incident of gaseous ion energy leads to the ejection of atom clusters [2]. In the sputtering process, sputtering gas is filled in an evacuated chamber. Free electrons bombard the gas particles at high voltage and produce gas ions. These charged ions initiate the electric field, and the continuous bombardment leads to the ejection from the surface target.

Thin layers of nanoparticles are efficiently deposited by annealing. The shape and size of nanoparticles are based on the thickness of the layer, substrate, heat, and time of annealing. Sputtering is cost-effective and with fewer impurities [3].

10.6.5 The Arc-Discharge Method

The arc discharge method described in Figure 10.10 is another synthesis method that helps in the production of nanostructured materials including fullerenes, carbon nano horns, and a few layers of graphene amorphous spherical-shaped nanoparticles. In the synthesis process, graphite rods are embedded in a chamber at pure helium pressure. Pure helium is significant to promote fullerene formation. Moisture and oxygen content inhibits

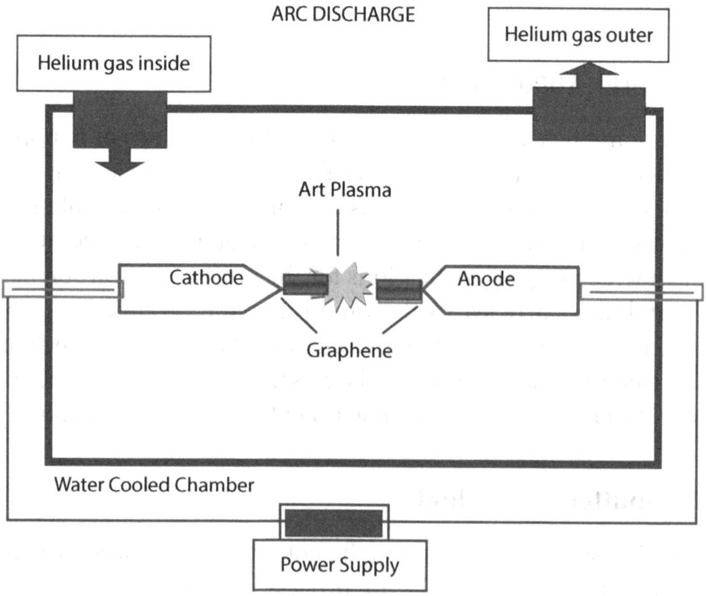

Figure 10.10 The arc-discharge method.

fullerene synthesis. Evaporation of carbon rods occurs by arc discharge between graphite rods [8].

10.6.6 Nanolithography Method

Nanolithography is another top-down method. The desired size and shape of particles can be obtained with the help of the lithography technique. Light-sensitive material is used for printing the favorable shape or structure of particles on a light-sensitive material. The material is removed and produces the desired particles. Highly complex instruments are required for this purpose. They are energy-requiring processes.

Various nanolithography mechanisms can be used for the preparation of electron beam, multiphoton, optical, scanning print, and nanoimprint lithography. Nanoimprint is a different lithography technique where soft polymeric material is stamped on a template material to form patterns. The template is made up of latex spheres. The limitation of the lithography method is the requisite of complex and costly equipment [13].

10.7 Conclusion and Future Scope

Recent advancements in nanomaterial synthesis have significantly transformed the field of nanotechnology. These developments have broadened potential applications in areas such as electronics, medicine, energy, and environmental management, allowing for the creation of materials with tailored properties. The ability to manipulate substances at the nanoscale has fundamentally changed traditional approaches and provided solutions to complex challenges.

Looking ahead, interdisciplinary collaboration will be essential for the future of nanomaterial synthesis. By integrating disciplines such as chemistry, materials science, and engineering, researchers can more effectively address the current limitations in synthesis techniques. This teamwork is expected to lead to more scalable and efficient production methods, enabling practical applications of advancements.

There is also a growing focus on sustainable practices, which is likely to drive innovations in green synthesis methods. These environmentally friendly approaches aim to reduce ecological impact while improving the performance of nanomaterials. As industries emphasize sustainability, the demand for greener production techniques will inspire research into new materials and processes that align with environmental goals.

Exploring multifunctional nanomaterials, such as hybrid and composite forms, offers promising research opportunities. These materials could provide enhanced functionalities and adaptability, leading to innovative applications in smart technologies and responsive systems. For instance, integrating various nanoscale components may produce materials that respond to environmental stimuli, opening new possibilities in sensing and energy conversion.

Additionally, effective collaboration between academic institutions and industry will be crucial for translating research into practical solutions. Such partnerships can speed up the development of applications, ensuring that cutting-edge research addresses real-world problems. As nanotechnology continues to evolve, nurturing these connections will be key to driving innovation and commercializing new nanomaterials.

In summary, the future of nanomaterial synthesis holds the potential to revolutionize technology and enhance the quality of life across various sectors. Ongoing investment in research, along with a strong emphasis on ethical considerations and regulatory frameworks, will be vital for the responsible development of nanotechnology. By prioritizing these aspects, the field is well-positioned to become a cornerstone of future scientific and technological progress, ultimately contributing to a more sustainable and innovative world.

References

1. Ayyub, P., Chandra, R., Taneja, P., Sharma, A.K., Pinto, R., Synthesis of nanocrystalline material by sputtering and laser ablation at low temperatures. *Appl. Phys. A Mater. Sci. Process.*, *73*, 67–73, 2001.
2. Son, H.H., Seo, G.H., Jeong, U., Shin, D.Y., Kim, S.J., Capillary wicking effect of a Cr-sputtered superhydrophilic surface on enhancement of pool boiling critical heat flux. *Int. J. Heat Mass Transfer*, *113*, 115–128, 2017.
3. Nie, M., Sun, K., Meng, D.D., Formation of metal nanoparticles by short-distance sputter deposition in a reactive ion etching chamber. *J. Appl. Phys.*, *106*, 054314, 2009.
4. Amendola, V. and Meneghetti, M., Laser ablation synthesis in solution and size manipulation of noble metal nanoparticles. *Phys. Chem. Chem. Phys.*, *11*, 3805, 2009.
5. Poolakkandy, R.R. and Menamparambath, M.M., Soft-template-assisted synthesis: a promising approach for the fabrication of transition metal oxides. *Nanoscale Adv.*, *2*, 5015–5045, 2020.

6. Martins, L., Alves Rosa, M.A., Pulcinelli, S.H., Santilli, C.V., Preparation of hierarchically structured porous aluminas by a dual soft template method. *Microporous Mesoporous Mater.*, *132*, 268–275, 2010.
7. Su, S.S. and Chang, I., *Commercialization of Nanotechnologies- A Case Study Approach*, pp. 15–29, Springer International Publishing, Cham, 2018.
8. Lieber, C.M. and Chen, C.-C., *Solid State Physics - Advances in Research and Applications*, vol. *48*, pp. 109–148, Academic Press, Cambridge, Massachusetts, 1994.
9. Meng, L.-Y., Wang, B., Ma, M.-G., Lin, K.-L., The progress of microwave-assisted hydrothermal method in the synthesis of functional nanomaterials. *Mater. Today Chem.*, *1–2*, 63–83. 68 Y, 2016.
10. Dong, Du, X., Liang, P., Man, X., One-pot solvothermal method to fabricate 1D-VS4 nanowires as anode materials for lithium ion batteries. *Inorg. Chem. Commun.*, *115*, 107883, 2020.
11. Bokov, D., Turki Jalil, A., Chupradit, S., Suksatan, W., Javed Ansari, M., Shewael, I.H., Valiev, G.H., Kianfar, E., Nanomaterial by Sol-Gel Method: Synthesis and Application. *Adv. Mater. Sci. Eng.*, *2021*, Article ID 5102014, 21, 2021. https://doi.org/10.1155/2021/5102014.
12. Shankar, S.S., Rai, A., Ahmad, A., Sastry, M., Biosynthesis of silver and gold nanoparticles from extracts of different parts of the Geranium plant. *Appl. Nanotechnol.*, *1*, 69–77, 2004.
13. Niederberger, M. and Pinna, N., *Metal Oxide Nanoparticles in Organic Solvents: Synthesis, Formation, Assembly and Application (Engineering Materials and Processes)*, Springer, Berlin, Germany, 2009.
14. Huang, J., Li, Q., Sun, D., Lu, Y., Su, Y., *et al.*, Biosynthesis of silver and gold nanoparticles by novel sundried Cinnamomum camphora leaf. *Nanotechnology*, *18*, 10, 104–105, 2007.
15. Li, H., Wang, L., Wei, Y., Yan, W., Feng, J., Preparation of Templated Materials and Their Application to Typical Pollutants in Wastewater: A Review. *Front. Chem.*, *10:*, 882876, 2022, doi: 10.3389/fchem.2022.882876.
16. Thomas, A. and Goettmann, F., Hard Templates for Soft Materials: Creating Nanostructured Organic Materials. *Chem. Mater.*, *20*, 3, 738–755, 2008, doi: 10.1021/cm702126j.
17. Christensen, L., Vivekanandhan, S., Misra, M., Mohanty, A., Biosynthesis of silver nanoparticles using murraya koenigii (curry leaf): An investigation on the effect of broth concentration in reduction mechanism and particle size. *Adv. Mat. Lett.*, *2*, 6, 429–434, 2011.
18. Bao, Y., Shi, C., Wang, T., Li, X., Ma, J., Recent Progress in Hollow Silica: Template Synthesis, Morphologies and Applications. *Microporous Mesoporous Mater.*, 227, doi: 10.1016/j.micromeso.2016.02.040.121-136,2016.
19. Liu, Y., Goebl, J., Yin, Y., Templated synthesis of nanostructured materials. *Chem. Soc. Rev.*, *42*, 2610–2653, 2013.
20. Liu, J., Yang, T., Wang, D.-W., Lu, G.Q., Zhao, D., Qiao, S.Z., A facile soft-template synthesis of mesoporous polymeric and carbonaceous nanospheres. *Nat. Commun.*, *4*, 2798, 2013.

21. Lv, R., Cao, C., Zhai, H., Wang, D., Liu, S., Zhu, H., Growth and characterization of single-crystal ZnSe nanorods via surfactant soft-template method. *Solid State Commun., 130,* 241–245, 2004.
22. Tang, T., Zhang, T., Li, W., Huang, X., Wang, X., Qiu, H., Hou, Y., Mesoporous N-doped graphene prepared by a soft-template method with high performance in Li–S batteries. *Nanoscale, 11,* 7440–7446, 2019.
23. Szczęśniak, B., Choma, J., Jaroniec, M., Major advances in the development of ordered mesoporous materials. *Chem. Commun., 56,* 7836–7848, 2020.
24. Wang, S. and Gao, L., Laser-driven nanomaterials and laser-enabled nanofabrication for industrial applications. *Ind. Appl. Nanomater.,* 181–203, 2019 Jan 1.
25. Spreafico, C., Russo, D., Degl'Innocenti, R., Laser pyrolysis in papers and patents. *J. Intell. Manuf., 33,* 353–385, 2022. https://doi.org/10.1007/s10845-021-01809-9.
26. Álvarez-Chimal, R. and Ángel Arenas-Alatorre, J., Green Chemistry for Environmental Sustainability—Prevention-Assurance-Sustainability (P-A-S) Approach. *IntechOpen,* 85–102, 2023. https://doi.org/10.5772/intechopen.1002203.
27. Baig, N., Kammakakam, I., Falath, W., Nanomaterials: A review of synthesis methods, properties, recent progress, and challenges. *Mater. Adv., 2,* 6, 1821–1871, 2021. https://doi.org/10.1039/D0MA00807A.
28. Eastoe, J., Hollamby, M.J., Hudson, L., Recent advances in nanoparticle synthesis with reversed micelles. *Adv. Colloid Interface Sci., 128–130,* 5–15, 2006. https://doi.org/10.1016/j.cis.2006.11.009.
29. Joy, J., Krishnamoorthy, A., Tanna, A., Kamathe, V., Nagar, R., Srinivasan, S., Recent Developments on the Synthesis of Nanocomposite Materials via Ball Milling Approach for Energy Storage Applications. *Appl. Sci., 12,* 18, 9312, 2022. https://doi.org/10.3390/app12189312.
30. Karunakaran, G., Sudha, K.G., Ali, S., Cho, E.-B., Biosynthesis of Nanoparticles from Various Biological Sources and Its Biomedical Applications. *Molecules, 28,* 11, 4527, 2023. https://doi.org/10.3390/molecules28114527.
31. Khan, N., Ali, S., Latif, S., Mehmood, A., Biological Synthesis of Nanoparticles and Their Applications in Sustainable Agriculture Production. *Nat. Sci., 14,* 06, 226–234, 2022. https://doi.org/10.4236/ns.2022.146022.
32. Li, F., Wan, Y., Chen, J., Hu, X., Tsang, D.C.W., Wang, H., Gao, B., Novel ball-milled biochar-vermiculite nanocomposites effectively adsorb aqueous As(V). *Chemosphere, 260,* 127566, 2020.
33. Lone, I.H., Radwan, N.R.E., Aslam, J., Akhter, A., Concept of Reverse Micelle Method For the Synthesis of Nano-Structured Materials. *Curr. Nanosci., 15,* 2, 129–136, 2019. https://doi.org/10.2174/1573413714666180611075115.
34. Manawi, Y.M., Ihsanullah, Samara, A., Al-Ansari, T., Atieh, M.A., A Review of Carbon Nanomaterials' Synthesis via the Chemical Vapor Deposition (CVD) Method. *Mater. (Basel Switzerland), 11,* 5, 822, 2018. https://doi.org/10.3390/ma11050822.

35. Mohan, S., Vellakkat, M., Aravind, A., . Hydrothermal synthesis and characterization of Zinc Oxide nanoparticles of various shapes under different reaction conditions. *Nano Express, 1*, 3, 030028, 2020. https://doi.org/10.1088/2632-959X/abc813.
36. Ndlwana, L., Raleie, N., Dimpe, K.M., Ogutu, H.F., Oseghe, E.O., Motsa, M.M., Msagati, T.A.M., Mamba, B.B., Sustainable Hydrothermal and Solvothermal Synthesis of Advanced Carbon Materials in Multidimensional Applications: A Review. *Materials, 14*, 17, 5094, 2021. https://doi.org/10.3390/ma14175094.
37. Peroulis, D., Waghmare, P.R., Mitra, S.K., Manakasettharn, S., Taylor, J.A., Krupenkin, T.N., Zhu, W., Nessim, G.D., Marano, F., Guadagnini, R., Rodrigues-Lima, F., Dupret, J.-M., Baeza-Squiban, A., Boland, S., Park, J.Y., Oh, S., Madou, M., Yap, Y.K., Aslan, B., Wannemacher, R., Chemical Vapor Deposition (CVD), in: *Encyclopedia of Nanotechnology*, Bhushan (Ed.), pp. 422–427, Springer Netherlands, 2012, https://doi.org/10.1007/978-90-481-9751-4_345.
38. Roberto Ruiz Hernández, A., Gutierrez Cruz, A., Campos-Delgado, J., Chemical Vapor Deposition Synthesis of Graphene on Copper Foils, in: *Graphene—A Wonder Material for Scientists and Engineers*, M. II, A.M., A. Bashir (Eds.), IntechOpen, London, United Kingdom, 2023, https://doi.org/10.5772/intechopen.106058.
39. Seong, G. and Rendón-Angeles, J.C. (Eds.), *Hydrothermal Synthesis of Nanoparticles*, MDPI, Basel, Switzerland, 2023, https://doi.org/10.3390/books978-3-0365-8065-4.
40. Singh, J., Dutta, T., Kim, K.-H., Rawat, M., Samddar, P., Kumar, P., 'Green' synthesis of metals and their oxide nanoparticles: Applications for environmental remediation. *J. Nanobiotechnol., 16*, 1, 84, 2018. https://doi.org/10.1186/s12951-018-0408-4.
41. Ying, S., Guan, Z., Ofoegbu, P.C., Clubb, P., Rico, C., He, F., Hong, J., Green synthesis of nanoparticles: Current developments and limitations. *Environ. Technol. Innovation, 26*, 102336, 2022. https://doi.org/10.1016/j.eti.2022.102336.
42. Savic, S., Vojisavljevic, K., Počuča-Nešić, M., Zivojevic, K., Mladenovic, M., Knezevic, N., Hard Template Synthesis of Nanomaterials Based on Mesoporous Silica. *Metall. Mater. Eng., 24*, (4), 2018, doi: 10.30544/400.
43. Babu, G.M.M. and Gunasekaran, P., Production and structural characterization of crystalline silver nanoparticles from Bacillus cereus isolate. *Colloids Surf. B Biointerfaces, 74*, 1, 191–195, 2009.
44. Jain, D., Kachhwaha, S., Jain, R., Srivastava, G., Kothari, S.L., Novel microbial route to synthesize silver nanoparticles using spore crystal mixture of Bacillus thuringiensis. *Indian J. Exp. Biol., 48*, 11, 1152–1156, 2010.
45. Gurunathan, S., Kalishwaralal, K., Vaidyanathan, R., Venkataraman, D., SR, P., *et al.*, Biosynthesis, purification and characterization of silver nanoparticles using Escherichia coli. *Colloids Surf. B Biointerfaces, 74*, 1, 328–335, 2009.

46. Sweeney, R.Y., Mao, C., Gao, X., JL, B., AM, B., *et al.*, Bacterial biosynthesis of cadmium sulfide nanocrystals. *Chem. Biol.*, *11*, 11, 1553–1559, 2004.
47. Sintubin, L., De Windt, W., Dick, J., Mast, J., van der Ha, D., *et al.*, Lactic acid bacteria as reducing and capping agent for the fast and efficient production of silver nanoparticles. *Appl. Microbiol. Biotechnol.*, *84*, 4, 741749, 2009.
48. Klaus, T., Joerger, R., Olsson, E., Granqvist, C.G., Silverbased crystalline nanoparticles, microbially fabricated. *Proc. Natl. Acad. Sci. U.S.A.*, *96*, 24, 13611–13614, 1999.
49. Zhang, H., Li, Q., Lu, Y., Sun, D., Lin, X., *et al.*, Biosorption and bioreduction of diamine silver complex by Corynebacterium. *J. Chem. Technol. Biotechnol.*, *80*, 3, 285–290, 2005.
50. Nanda, A. and Saravanan, M., Biosynthesis of silver nanoparticles from Staphylococcus aureus and its antimicrobial activity against MRSA and MRSE. *Nanomedicine*, *5*, 4, 452–456, 2009.

11

Emerging Trends and Future Developments in Smart Materials and Their Applications: A Comprehensive Review

S. Grace Infantiya[1*], D. Anbuselvi[2] and N. Suthanthira Vanitha[3]

[1]Department of Physics, Sri Sairam Institute of Technology, Chennai, Tamil Nadu, India
[2]Department of Physics, T. John Institute of Technology, Bangalore, India
[3]Department of Electrical and Electronics Engineering, Muthayammal Engineering College (An Autonomous Institution), Namakkal, India

Abstract
The advancement of material science and nanotechnology has materialized as a transformative field, with widespread implications in an assortment of science and industrial realms. Within this panorama, technologically advanced (smart) nanomaterials—engineered materials with tailored attributes—are in the limelight of innovations. This proposed chapter portrays an extensive overview of smart nanomaterials, emphasizing their importance in material science and ground-breaking applications.

Smart nanomaterials are distinguished by their unique ability to adapt dynamically to external stimuli, including temperature, pH, light, electromagnetic radiation, and mechanical stresses. Smart nanomaterials can alter their chemical and physical attributes in anticipation of particular stimuli, making them extremely adaptable and advantageous in an array of applications, from targeted drug delivery systems in biomedicine to adaptive materials in electronics and energy storage. The incorporation of smart nanomaterials into multiple technologies symbolizes an important breakthrough in the way we approach and resolve multifaceted challenges. Their nanoscale magnitude promotes a high surface area-to-volume ratio, facilitating sensitivity and interaction with other kinds of material. Furthermore, the potential to optimize their

Corresponding author: graceinfantiya@gmail.com

Kolla Bhanu Prakash, S. V. Ranganayakulu and K. S. Jagannatha Rao (eds.) New Frontiers in Materials Science: Interdisciplinary Approaches to Innovation and Technologies, (215–240) © 2025 Scrivener Publishing LLC

characteristics at the atomic or molecular level enables the development of materials with specialized functionality. This accuracy in design leads to momentous enhancements in performance and proficiency across various applications.

The chapter unveils an overview of the basic concepts of nanotechnology and materials science, laying the groundwork for comprehending the unique characteristics of smart nanomaterials. It examines the development of nanotechnology, highlighting major milestones and achievements that have paved the path for contemporary advances. Following this, the chapter discusses the synthesis and characterization of smart nanomaterials. It includes sophisticated manufacturing techniques including self-assembly, sol–gel processes, electrospinning, and lithography. These techniques provide exact control over the size, shape, and composition of smart nanomaterials, which is critical for designing their responsive behaviors.

The primary focus of this chapter is on the different implementations of smart nanomaterials across several fields:

Biomedical Applications: Smart nanomaterials are transforming healthcare through advances in medication delivery, allowing them to target specific cells or tissues, minimizing adverse effects and increasing therapeutic effectiveness. They are also employed in diagnostic imaging and biosensors, which allow for real-time illness monitoring and diagnosis.

Environmental Applications: Smart nanomaterials are utilized in pollution monitoring and water, as well as air purifications. Their capacity to absorb and the decompose harmful substances at the molecular level makes them advantageous in resolving ecological problems and fostering sustainability.

Energy Applications: Smart nanomaterials are employed in the energy industry to fabricate more effective and robust solar cells, batteries, and supercapacitors. Their particular features enable more efficient energy storage and conversion, which is critical for the advancement of renewable energy technology.

Electronic and Smart Technologies: The incorporation of smart nanomaterials into electronic gadgets promotes the production of versatile, portable, and responsive electronics. These materials improve the efficiency and functioning of devices such as actuators, sensors, and other components used in smart technology and the Internet of Things (IoT).

The chapter eventually ends with an examination of the future potential and challenges associated with smart nanomaterials. It examines the ethical and environmental implications of nanotechnology, highlighting the significance of responsible research and application.

This chapter intends to stimulate future research and innovation in the rapidly evolving field of smart nanomaterials by offering an unambiguous understanding of their enormous potential. It is an invaluable resource for researchers, technologists, and students, providing insights into the concepts, applications, and potential futures associated with intelligent (smart) nanomaterials.

To summarize, smart nanomaterials are a vital junction of material science and nanotechnology, providing unparalleled prospects for expansion across multiple industries. This chapter emphasizes their significance, the novel uses they promote, and the future opportunities they promise for research and industry.

***Keywords*:** Smart materials, nanomaterials, sensors, biomedical applications, CNT's, shape memory polymers, photo-sensitive polymers

11.1 Introduction

Over the course of time, nanotechnology and nanomaterials have contributed significantly to the advancement of scientific and technological domains. Nanomaterials, specifically graphene, nanotubes, metal nanoparticles, and polymer nanoparticles, have received abundant investigation. Nanomaterials with small dimensions ($\leq$ 100 nm) have exceptional physical and chemical attributes, such as higher molar extinction coefficients, more effective reactivity, enhanced absorption and surface area, flexible plasmonic attributes, distinctive photo and magnetic characteristics, and quantum phenomena due to their relatively small dimensions [1, 2]. As a result, several types of nanomaterials have been effectively investigated and employed in an assortment of industries, spanning healthcare, electronic devices, energy, and the environment [3].

In parallel with the advent of advanced nanomaterials and sophisticated miniaturized devices pivotal in enriching human health and the excellence of life, smart materials are at the cutting edge of this transformation [2]. These materials are remarkable in adapting dynamically to environmental stimuli such as temperature, pH, gas, pressure, light, electric or magnetic fields, and chemical conditions. Due to their exceptional properties of stimulus sensitivity and autonomous behavior, smart materials serve as the underpinning for numerous technological spreads. Conventional examples include chemical and strain sensors, piezoelectric compounds, actuators, optoelectronic devices, electronic switches, artificial muscles, and controlled drug delivery. These advancements have paved the way for cutting-edge research that challenges standard design approaches and substantially expanded application scope [1–6]. For example, robotics is advancing from heavy body frameworks to lightweight robots made of flexible, adaptable substances. Futuristic automated systems will require a cohesive combination of mechanical engineering, materials science, chemistry, and biotechnology since the robots can expand, regenerate, and transform their appearance and functionality in interaction with their physical and chemical environments.

Smart materials additionally serve as the foundation for cutting-edge technologies that merge the boundaries between living and non-living entities by allowing artificial structures to effortlessly communicate with biological species. Additionally, from the conventional applications of environmental sensors and actuators, smart materials are propelling notable improvements in physiological sensing, robotic surgery, targeted medication administration, human-computer interface, and rehabilitation medical technology [6–8]. They are also vital for the evolution of soft robotics and bioelectronics. The next initiation of smart materials will offer enhanced functionality, superior controllability, and improved biocompatibility, aiding a future where technological and biological systems converge to create unprecedented solutions for human advancement.

11.2 What are Smart Materials

Smart materials are a fascinating category of materials that may change their characteristics by interacting with external stimuli. These materials have an extraordinary ability to change structure, density, texture, color, modulus, stiffness, heat generation, hyperthermia, magnetic effects, optical properties, surface area, solubility, and mechanical deformation in response to external physical and chemical or biological stimuli including temperature, ultrasound, light, pressure, electric or magnetic fields,

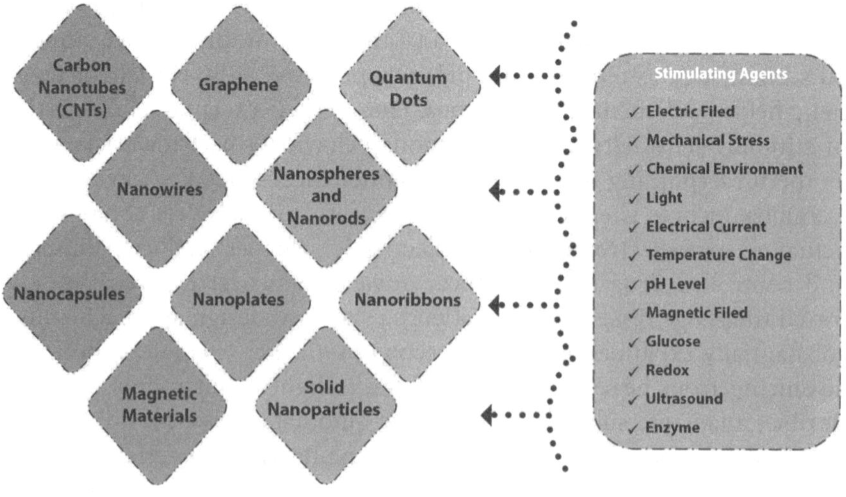

Figure 11.1 Schematic illustration of nanomaterials and their associated stimulating factors.

Table 11.1 Various nanomaterials and their corresponding stimuli, and potential applications.

Nanomaterials	Stimulating agent	Responses	Applications
Carbon Nanotubes, and Graphene	Mechanical stress, electric field, and chemical environment	Modifications in electrical conductivity and mechanical flexibility	Energy storage, composite materials, sensors, and flexible electronics [6, 7]
Quantum Dots	Wavelength (light)	Fluorescence emissions and variations in the optical characteristics	Electronic displays, solar cells, and biological imaging [2]
Nanowires	Electrical current and temperature	Modifications in electrical resistance and thermal conductivity	Biosensors, electronic devices, and energy harvesting [12]
Nanospheres and Nanorods	Redox and pH variations	Optical properties	Biological imaging, drug delivery, and catalysis [12]
Nanocapsules	Variations in pH, temperature, chemical	Release of encapsulated compounds, deviations in solubility	Targeted drug transport, controlled release systems [11]
Nanoplates	Temperature, mechanical stress	Modifies surface properties and mechanical flexibility	Sensors and flexible wearable electronics [9]
Nanoribbons	Electrical fields, mechanical pressure	Modifications in mechanical and electrical characteristics	Electronic gadgets and sensors [1]

(*Continued*)

Table 11.1 Various nanomaterials and their corresponding stimuli, and potential applications. (*Continued*)

Nanomaterials	Stimulating agent	Responses	Applications
Magnetic Nanoparticles	Magnetic fields	Modifications in magnetic properties and hysteresis	Data storage, memory devices, MRI, and hyperthermia therapy
Graphene Oxide (GO)	Redox and pH	Modifications in optical and electrical conductivity	Water purification and filtration, biological sensing [7]
Metallic nanoclusters	Redox, wavelengths	Modifications in optical absorption and fluorescence	Sensing devices, bioimaging, and catalysis [13]
MXenes (2D)	Electric Fields, mechanical pressure	Modifies mechanical and electrical properties	Electromagnetic shielding, energy storage/ conversion, and sensing devices [13, 14]

chemical constraints, pH, solvent, electrochemical, adenosine triphosphate (ATP), H_2O_2, enzyme, redox potential, and hypoxia [9, 10]. Figure 11.1 depicts a schematic illustration of nanomaterials and their associated stimulating factors.

For instance, "smart" materials may exhibit volumetric changes such as expansion or shrinkage, undergo chromatic transitions by altering their color, experience variations in mechanical properties such as stiffness or elasticity, or undergo reversible shape transformations. This dynamic adaptability is driven by their intrinsic ability to respond to external stimuli, making them highly versatile for advanced applications, from medical devices and smart textiles to advanced sensors and adaptive structures, promising continued innovation and enhanced functional capabilities in the future [11, 15]. Table 11.1 illustrates how multiple nanomaterials respond to diverse stimuli, as well as their prospective applications highlighting their adaptability and usefulness in technological advancement.

11.3 Smart Materials vs. Conventional Materials

Smart materials are distinct from conventional substances in terms of their intrinsic traits and flexibility to external stimuli. Advanced materials are optimized to demonstrate precise, often reversible, modifications to their properties in response to external stimuli, while traditional materials typically maintain a fixed array of characteristics under multiple conditions [14].

11.3.1 Properties and Functionalities

11.3.1.1 Conventional Materials

Traditional substances, especially metals, ceramics, and conventional polymers, are distinguished according to their static characteristics. Their mechanical, thermal, and chemical characteristics remain near-constant unless subjected to extreme conditions.

- **Metals:** Metals most commonly show constant strength, toughness, and thermal conductivity. Their attributes are controlled by their composition and microstructure, and they remain unaltered in adaptation to environmental changes under typical circumstances [1, 2].
- **Ceramics:** Ceramics are renowned for their hardness and high-temperature stability, and their properties fluctuate minimally when exposed to external stimuli, making them ideal for high-temperature applications.
- **Polymers:** Classical polymers possess permanent mechanical attributes and thermal stability, which do not respond to external stimuli. They are commonly employed in applications that demand predictable behavior without the necessity for dynamic reactions [14].

11.3.1.2 Smart Materials

Smart materials are intended to automatically react to certain external factors by altering their properties in a controlled and frequently reversible manner [2, 9, 13, 15].

- **Shape Memory Alloys (SMAs):** They exhibit a phase alteration between martensite and austenite and are capable of returning to their original shape after deformation when exposed to fluctuations in temperature. This capacity is utilized in applications that require form recovery and actuation [16].
- **Piezoelectric Materials:** These compounds fabricate an electrical charge in response to mechanical stress, and vice versa. They can be widely utilized in sensors and actuators [17, 18].
- **Shape Memory Polymers (SMPs):** These compounds are characterized by adjustable phase transitions that allow them to return to a predefined structure when exposed to specific stimuli such as temperature or light [17].
- **Thermochromic Materials:** These materials transform shade in accordance with temperature deviations attributable to modifications in their molecular framework, impacting light absorption and reflection attributes.
- **Nanomaterials:** Carbon nanotubes (CNTs) and graphene have incomparable mechanical and electrical properties that can be exploited in a wide variety of advanced applications, including flexible electronics and high-strength composites.

11.3.2 Sensitivity and Adaptability

11.3.2.1 Conventional Materials

They showcase restricted versatility to environmental factors. Their mechanical and physical features are predetermined and are not affected by adaptation to environmental conditions. For example:

- **Structural Integrity:** Despite conventional metals and ceramics maintaining their structural stability under normal circumstances, they may fail to operate effectively when subjected to exceptional or dynamic pressures.
- **Thermal Stability:** While ceramics extend high-temperature equilibrium, they do not change with modifications in temperature beyond their constructed limitations, often resulting in fracture or degradation.

11.3.2.2 Smart Materials

Smart materials are recognized by their capacity utilization to dynamically adapt and respond to external stimuli, thereby enhancing effectiveness and diversity. For example:

- **Dynamic Response:** Shape memory alloys and shape memory polymers may experience substantial modifications in shape or mechanical characteristics in response to temperature fluctuations, providing alternatives for applications that require adaptability and recovery.
- **Multi-Stimuli Interaction (SIM):** Certain smart materials may respond to various types of stimuli (e.g., temperature, light, and electric fields) simultaneously, enabling multifunctional capabilities that are not achievable with ordinary materials [15].

11.4 Classifications of Smart Materials

Traditional smart materials are primarily based on metal alloys, inorganic compounds, polymers, and carbon-based nanomaterials. Smart nanomaterials are categorized into subclasses based on their responses to distinct stimuli. External parameters such as mechanical stress, temperature, moisture, pH, and electric as well as magnetic fields can precisely influence the properties of smart materials [17–19].

Shape memory alloys (SMAs) and piezoelectric ceramics are typically utilized in the assembly of actuators due to their proficiency in responding to external stimuli by changing their shape or generating electrical signals [14]. However, their real-world applications are frequently constrained by the requirement for challenging working environments and/or considerable operating voltages. Polymer-based smart materials, on the other hand, have advantages such as flexible properties, lightweight construction, and high transparency, making them ideal for applications requiring flexibility and aesthetics. Despite these advantages, they have challenges such as low mechanical durability, slower response time, and insufficient environmental stability [19]. Carbon nanoparticles constitute an intriguing alternative due to their distinct characteristics. They are generally accessible, stable, and have highly effective mechanical strength and flexibility. Furthermore, carbon nanoparticles have high electrical and thermal conductivity. These features make them ideal candidates for the development of new smart materials and electronic devices with high-performance possibilities. Their combination of affordability, stability, and functionality addresses many of the shortcomings of traditional smart materials, paving opportunities for emerging applications in an assortment of domains [7]. Carbon nanotube (CNT)-based electromechanical actuators, for instance, are capable of massive strain deformations at low applied voltages, highlighting their potential for implementation in smart materials. However, the development of altering or manipulating CNTs to improve their 'smart' features can occasionally be expensive and complex. In comparison, graphene-based smart materials have significant advantages due to their huge specific surface areas, strong electrical and thermal conductivities, and excellent mechanical attributes. Chemically modified graphene (CMG) compounds, such as graphene oxide (GO), reduced graphene oxide (rGO), and its derivatives, can be fabricated inexpensively and on a large scale [20, 21]. These materials can be easily processed into various configurations, including fibers, films, and porous frameworks, and can be further modified or hybridized to enhance their functionalities. The versatility and cost-effectiveness of graphene and its derivatives make them highly promising for the development of high-performance smart materials, capable of surpassing some of the limitations faced by traditional materials like CNTs.

There are various types of smart materials available based on their characteristics, a few of them are explained below.

11.4.1 Shape Memory Polymers (SMPs)

Shape memory polymers (SMPs) are highly sophisticated smart materials that can be flexed and restricted into a temporary configuration before reverting to their original arrangement when an external stimulus is released. These materials stand out for their structural adaptability, lightweight nature, affordable cost, ease of manufacturing, and capacity to withstand significant elastic strain (above 200%). Additionally, SMPs are biocompatible and biodegradable, making them advantageous for a variety of applications [22–24].

11.4.1.1 Stimulation Mechanisms for Shape Memory Polymers

Shape memory polymers (SMPs) function through numerous stimulation mechanisms that allow them to revert to their original shape upon the application of specific external stimuli, including light, temperature, and electric field. This ability is fundamentally linked to their molecular architecture, which comprises net-points and switching domains [22, 23].

1. Molecular Architecture:

Net-Points: Net-points are the predefined cross-linking points in the polymer network. They maintain structural stability and indicate the SMP's definitive shape. Net-points are typically created by chemical cross-linking or physical interactions (such as hydrogen bonds or ionic interactions) that strengthen the polymer matrix.

Switching Domains: These are the regions within the polymer that undergo reversible phase transitions. They are responsible for the temporary shape change and recovery process. Switching domains can be thermally activated (as in thermoplastic SMPs), light-sensitive (as in photo-responsive SMPs), or electro-responsive (as in electro-active SMPs). Figure 11.2 shows the schematic representation of Molecular architecture for stimulation of SMPs.

2. Stimulation Mechanisms:

- **Thermal Stimulation:**

Mechanism: When SMPs undergo thermal expansion above their transition temperature (T_{trans}), the switching domains switch from glassy or crystalline to rubbery or amorphous. This phase transition enables the polymer to adopt a new temporary arrangement [26]. When the material undergoes cooling, it " twists " into this shape, and when heated again, it returns to its initial, permanent configuration (Figure 11.2).

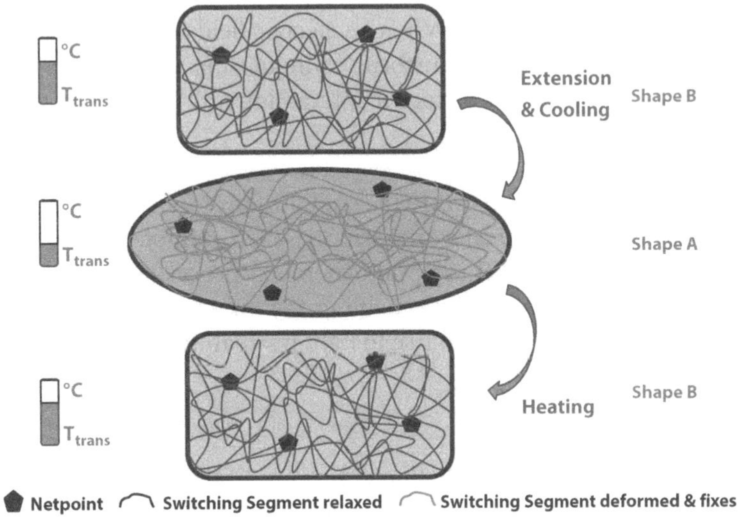

Figure 11.2 Molecular mechanism for stimulation of shape memory effect in SMPs.

Types:

i. **Thermoplastic SMPs:** When subjected to heat and cooled, these compounds undergo a reversible transformation from being soft, flexible to a hard, rigid phase.
ii. **Thermosetting SMPs:** When heated, these polymers experience a chemical reaction that solidifies their molecular structure, with reversible transitions achievable in certain circumstances.

Example: Polyurethane-based SMPs frequently employ this process, in which the transition temperature is precisely tailored to suit their particular applications desires.

- **Electric Filed Stimulation:**

Mechanism: In electro-responsive SMPs, an electric field modifies the ionic or dipole interactions in the switching domains. This modification can lead the polymer to break down or reform to its original shape in response to an electrical charge being applied [27].

Types:

i. **Conductive SMPs:** These have fillers that are conductive or are intrinsically conductive, allowing them to adapt to electrical currents.
ii. **Electroactive SMPs:** When an electric field is applied, these kinds of polymers transform configuration due to electrostatic attraction or dielectric interactions.

- **Light Activation:**

Mechanism: Photo-responsive SMPs contain photo-sensitive compounds that modify their chemical composition when stimulated by particular wavelengths of light. These photochemical processes transformation impacts the switching domains, enabling the polymer to undergo deformation or regain shape in response to light irradiation [22].

Types:

i. **Photochromic SMPs:** Photochromic SMPs are composed of up of photochromic substances that modify their appearance and form in response to ultraviolet light (UV).

ii. **Photothermal SMPs:** Photothermal SMPs absorb photons and transmit it to heat, allowing polymer molecules to change phase and its structure.

Example: SMPs treated with azobenzene molecules, which undergo photoisomerization, can change shape when exposed to UV or visible light.

- **Magnetic Stimulation:**

Mechanism: Magnetic-responsive SMPs are composed of particles with magnetic properties that interact to an external magnetic field. The magnetic particles produce specific heating or magnetic forces, causing modifications in shape in the polymer [22].

Types:

i. **Magneto-Responsive SMPs:** These incorporate magnetic nanoparticles (such as iron oxide) that may be influenced by external magnetic impulses to cause structural transformations.
ii. **Magnetothermal SMPs:** These harness magnetic field-induced heating to accomplish phase transitions and shape restoration.

- **Chemical Stimulation:**

Mechanism: Chemically sensitive SMPs influence their physical characteristics in response to certain chemical influences, such as pH switches or exposition to chemicals. The chemical interactions cause inflammation, shrinking, or additional structural alterations, resulting in morphological modification [22].

Types:

i. **pH-Sensitive SMPs:** These synthetic polymers transform in shape or volume in response to fluctuations in pH, making them suitable for therapeutic administration and environmental monitoring.
ii. **Solvent-Sensitive SMPs:** These alter their configuration when exposed to chemical solvents, resulting in swelling or deswelling effects.

- **Humidity Stimulation:**

Mechanism: Humidity-sensitive SMPs alter their shape in accordance to moisture conditions. The absorption or desorption of water causes expansion or compression, resulting in form changes.

Type:

i. **Hydrogel-Based SMPs:** These polymers hatve hydrophilic structures that accumulate water, resulting in swelling and structural changes in response to fluctuations in moisture.

11.4.1.2 Applications of Shape Memory Polymers

Shape memory polymers (SMPs) are exceedingly multipurpose and advantageous compounds due to their lightweight architecture, cost-effectiveness, relatively simple manufacturability, biodegradability, and biocompatibility. Their exceptional mechanical features, involving their excellent elasticity as well as their ability to withstand notable structural alterations, further boost their popularity. As a result, SMPs are used in a wide variety of applications (Figure 11.3). They are utilized in medical instruments like self-expanding stents, orthodontic cables, and lightweight surgical instruments due to their biocompatibility as well as capability to reset to a predefined configuration. In the aviation and automobile sectors,

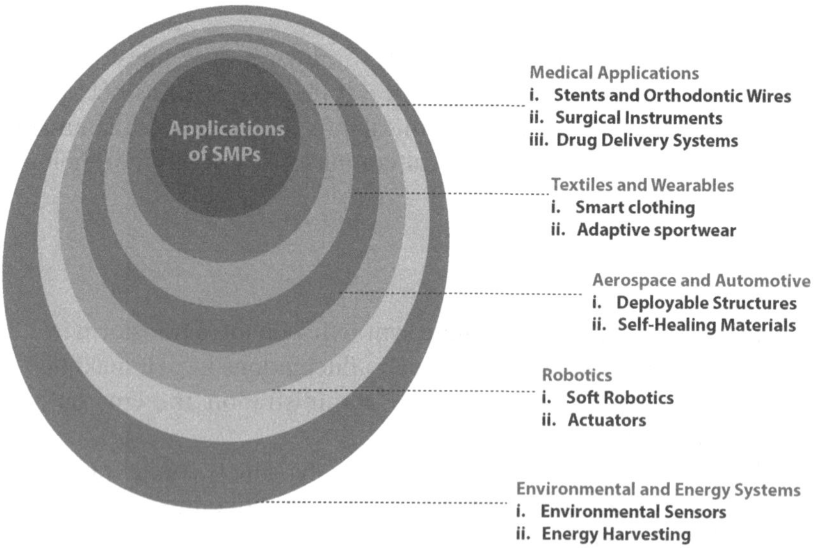

Figure 11.3 Schematic demonstration of various applications of shape memory polymers.

SMPs are implemented to develop flexible frameworks and actuators that accommodate to fluctuating operational environments. SMPs are also employed in intelligent fabrics and portable technology because of their tendency to adapt to shape in response to external stimuli, resulting in both convenience and functionality. Their possible applications include environmental sensors, pharmaceutical distribution networks, and self-healing components, demonstrating their wide scope and profound influence across multiple technical areas [23–27].

11.4.2 Photo-Sensitive Polymers (PSPs)

Photo-sensitive polymers (PSPs) are cutting-edge functional materials that undergo reversible changes in their physicochemical properties in response to light (photon) stimuli. These materials are integral to numerous high-tech applications, including smart coatings and sensors [28]. The types and operational principles of photo-responsive polymers, as well as their applications, are detailed below.

11.4.2.1 Types and Operational Principles

- **Photochromic Polymers**

Principle: When photochromic polymer compounds are subjected to light, they demonstrate irreversible variations in color or absorption characteristics. These alterations are triggered by structural or electrical modifications in photochromic molecules, which often involve transitions between distinct isomers or oxidation conditions [28, 29].

Examples: Photochromic polymers integrating spiropyrans, azobenzenes, or diarylethenes change their configuration when exposed to light. When illuminated by UV or visible light, these molecules change their configuration, resulting in a color shift or change in optical properties.

- **Photothermal Polymers**

Principle: Photothermal polymers absorb light, particularly in the infrared or visible spectrum, and transfer it to heat energy. This concentrated heating may trigger transitions between phases, such as melting or swelling, resulting in macroscopic changes in the polymer's shape or characteristics. Examples of photothermal nanoparticles embedded within polymer compounds include carbon nanotubes and gold nanoparticles. When exposed to light, these nanoparticles heated up, causing the polymer matrix to alter form or functionalities as required [30].

- **Photomechanical Polymers**

Principles: Photomechanical polymers change their mechanical characteristics or dimensions in response to light. This is commonly accomplished by light-induced phase transitions or interactions between light-sensitive chemicals and the polymer matrix.

Examples: Polymers containing azobenzene or other photo-responsive compounds that exhibit photo-induced mechanical deformations or extensions when exposed to specific wavelengths of illumination [31].

- **Photo-Sensitive Polymers with Self-Motivated Covalent Bonds**

Principle: These polymers have dynamic covalent interactions that can be reconfigured or fractured by light exposure, causing modifications in the polymer's arrangement or attributes. This type of polymer can undergo both reversible and irreversible alterations in response to light-induced chemical processes.

Example: Photo-cleavable or photo-cross-linkable polymer compounds experience significant structural modifications when exposed to light [32].

11.4.2.2 Applications in Smart Coatings and Sensors

11.4.2.2.1 Smart Coatings

i. **Photochromic Coatings:** These coatings can adapt coloration in combination with light, making them suitable for installation on sophisticated windows and coloration-changing surfaces. Photochromic coatings, for example, may modify color dynamically to block or transmit light, increasing energy efficiency and serving as visual indicators in a variety of applications.
ii. **Photothermal Coatings:** Photothermal coating compounds are used in applications that need heat production or dissipation. Photothermal coatings, for example, can be utilized to create self-cleaning surfaces or in applications that require targeted heating for restoration or stimulation.

11.4.2.2.2 Sensors

i. **Photo-Responsive Sensors:** These types of sensors incorporate photo-responsive monomers to recognize and respond to fluctuations in light intensity, wavelength, or polarization.

They are helpful for environmental monitoring, optical communication, and biomedical diagnostics.

ii. **Photochromic Sensors:** These sensors can be utilized in operations that require visual signals for recognizing the presence of particular triggers or situations. The chromatic change of photochromic polymer compounds provides a simple approach to monitor fluctuations in the environment.

These types of photo-responsive polymers and their associated applications highlight the versatility and innovation potential in smart coatings and sensors, driven by advances in material science and light-responsive technologies [32, 33].

11.4.3 Carbon Nanotubes (CNTs)

Carbon-based nanomaterials have attracted extensive attention in real-world implications because they are widely accessible, have minimal synthesis costs, and are simple to synthesize. These nanoparticles are classified in a multitude of methods that range reflecting their broad spectrum of size, shape, and chemical composition. This variety enables the specialized design and optimization of materials to satisfy specific application demands [34–36]. Carbon-based nanomaterials are categorized primarily based on their microstructure and physical and chemical characteristics. These materials are commonly classified as carbon quantum dots (CQDs), fullerenes, carbon nanotubes (CNTs), graphene and its derivatives, nanodiamonds, and graphene oxide (GO). Each class has distinctive physicochemical features that determine its behavior and applications [7, 35].

Carbon-based stimulation-responsive components, in particular, stand out for their sensitivity and their capability to profoundly alter their behavior in response to certain external stimuli. These materials are exceptionally chemically resistant and have a wide range of functions [37]. Surface functionalization improves its efficiency and effectiveness by allowing for precise modifications to satisfy specified needs. Furthermore, carbon-based stimuli-responsive compounds have the advantageous effects of chemical alterations, which addresses genetic compatibility difficulties in cancer therapy and optimizes their effectiveness as drug carriers [38]. These materials can act as smart networks with dynamically flexible physicochemical properties, responding automatically to modifications in internal as well as external environmental stimuli. The diversity of nanostructure combinations and molecular designs, combined with their incorporation into functional complexes including diverse carriers, opens up novel possibilities for the fabrication of highly sophisticated smart nanomaterials [39].

11.4.3.1 Properties

- **Mechanical Strength:** CNTs exhibit excellent mechanical strength and stiffness owing to their distinct tubular nanostructure and sp^2-hybridized carbon bonds, making them suitable for strengthening substances and composites.
- **Electrical Conductivity:** CNTs have remarkable electrical properties, which can be tuned to various kinds of electronic applications. This characteristic is caused by the delocalized π-electrons along the nanotube walls, which enable them to conduct electricity efficiently.
- **Thermal Conductivity:** CNTs have high thermal conductivity, which is advantageous for applications requiring efficient heat dissipation or thermal management.
- **Stimulus Response:** CNTs can be personalized to respond to external influences such as temperature, light, magnetic fields, and chemical environments. This responsiveness is frequently achieved by incorporating specific functional groups or integrating other responsive materials [34–39].

11.4.3.2 Applications of Carbon-Based Stimuli-Responsive Materials

Carbon-based stimuli-responsive constituents, such as carbon quantum dots (CQDs), carbon nanotubes (CNTs), graphene, fullerenes, nanodiamonds, and graphene oxide and its derivatives, have emerged as fundamental elements in advanced technologies due to their exceptional properties and ability to respond to various stimuli [39, 40]. Carbon-based stimuli-responsive materials are extremely responsive and can experience substantial behavioral changes. They are chemically stable and have considerable potential for a variety of applications. These materials' surfaces can be modified to respond to various environmental elements such as magnetic fields, temperature, pH, light, and humidity. These stimuli cause changes in hydrogen bonding and interactions between functional groups, increasing conductivity and other physical and chemical characteristics. Furthermore, their ability to undergo chemical alterations resolves issues like genetic compatibility in cancer therapy, making them more effective drug carriers. The diverse spectrum of nanostructure combinations, molecular designs, and functional complexes with different carriers opens new avenues for producing superior smart nanomaterials.

Several research studies have demonstrated carbon-based stimuli-responsive nanoparticles' efficacy in illness detection and therapy. These materials have shown great promise, especially in cancer therapy. As a result, many researchers are working on integrating carbon-based nanomaterials with other functional materials to provide novel platforms for disease diagnosis and therapy. For example, Zhang *et al.* [2] developed stimuli-responsive nanomaterials by integrating carbon dots with fluorescent and photoacoustic imaging capabilities, as well as photothermal and photodynamic cancer treatments. These nanoparticles were incorporated into gels that detect glutathione and pH levels. Stimulated by the tumor microenvironment (TME), the composite eventually degrades, allowing the material to be released.

- **Electronics and Sensors**

Flexible Electronics: Carbon-based materials, notably CNTs and graphene, are commonly utilized in flexible electronic components because of their mechanical adaptability, excellent electrical conductivity, and lightweight characteristics. They allow for the fabrication of flexible, adaptable, and lighter electronic devices such as flexible monitors, wearable technology, and sophisticated textiles.

Sensors: Carbon-based materials that react to stimulation are vital in sophisticated sensor technology. Their huge surface area and conductivity make them ideal to recognize various kinds of environmental stimuli such as gases, humidity, temperature, and light. Modified carbon nanotubes and graphene can improve sensor sensitivity and response times [40–42].

- **Energy Storage and Conversion**

Supercapacitors: Carbon-based materials are employed in supercapacitors due to their substantial surface area and conductivity properties, which are critical to improving energy storage and charge/discharge cycles. CNTs and graphene-based supercapacitors provide high energy density and quick charging.

Battery Technological Advances: Carbon-based materials have been employed in batteries to enhance conductivity and electrode efficiency. Their function in lithium-ion and other modern battery technologies is improving battery lifespan and durability [43, 44].

CNTs are utilized as electrically conducting additions in lithium-ion batteries to improve electrode conductivity and battery performance.

- **Environmental and Water Purification**

Water Purification: Carbon-derived compounds, including activated carbon and graphene oxide, are utilized to detoxify and treat water. Their large surface area and ability to absorb contaminants make them ideal for eliminating organic compounds, toxic substances, and other impurities from water.

Pollution Control: Carbon nanotubes and graphene are used for ecological management and pollution management systems. They can absorb contaminants and serve as catalysts for chemical reactions that neutralize hazardous substances. For instance, CNT-based filters and adsorbents are used to collect airborne contaminants and eradicate them from emissions from industrial processes [45, 46, 49].

- **Probes**

Carbon-related stimuli-responsive compounds have gained popularity as fluorescent probes due to their outstanding sensitivity and specificity for recognizing a variety of chemical substances. These materials, which include carbon nanotubes and graphene derivatives, are especially good for detecting metal ions in water samples, with great precision and dependability. Besides metal ion identification, they have tremendous promise for sensing anions, organic molecules, and other chemical entities. This versatility creates new opportunities in a variety of key industries, including food safety, chemical manufacturing and environmental preservation, where accurate and sensitive detection of pollutants and analytical substances is critical to ensuring compliance as well as safety [34, 47].

- **Bioimaging**

Carbon-based stimuli-responsive materials are increasingly utilized in bioimaging due to their superior optical properties and biocompatibility [51]. These materials, such as graphene and carbon nanotubes, enhance imaging techniques by providing high resolution and contrast. Their ability to be functionalized for targeted imaging allows for detailed visualization of cellular structures and processes. Additionally, their strong photoluminescence and photothermal properties facilitate advanced imaging applications, including fluorescence and thermal imaging, offering significant improvements in the accuracy and depth of biological observations [39, 48–50].

11.5 Further Research Insights

As the discipline of smart materials continues to develop, more study into new trends provides chances to maximize their potential. The development of multi-functional smart materials, such as stimuli-responsive, carbon-based, and bioinspired materials, is especially fast, pushing the frontiers of applications in domains such as healthcare, energy, and environmental sustainability.

i. Nanostructure Implementation and Surface Functionalization
One important topic for future study is the combination of nanostructures with customized surface functionalization. These methods can improve smart materials' sensitivity and reactivity to environmental stimuli including temperature, pH, magnetic fields, light, and humidity. Future studies might look at more efficient techniques for changing the surface characteristics of materials to improve their performance in a variety of applications, including drug delivery, energy storage, and biosensing.

ii. Biomedical Applications and Personalized Medicine
In the biomedical arena, smart materials, particularly stimuli-responsive nanoparticles, hold promise for individualized therapy. Carbon-based nanomaterials, such as carbon dots, graphene, and nanotubes, have demonstrated promise in diagnostics, targeted drug administration, and less invasive therapies, notably for cancer therapy. Further research should concentrate on addressing biocompatibility issues, improving drug release mechanisms in response to the tumor microenvironment, and combining imaging and therapeutic functions into a single platform for theranostics.

iii. Sustainable and Green Synthesis Approaches
With a growing emphasis on sustainability, the creation of environmentally friendly smart materials is critical. Green synthesis strategies, such as using plant-based or microbial processes to manufacture nanoparticles, show great potential. Green synthesis can limit the use of harmful chemicals, reducing environmental impact and improving nanomaterial biological compatibility. Further research should focus on improving and scaling these processes for commercial applications, notably in antimicrobial treatments, environmental remediation, and renewable energy technologies.

iv. Smart Materials in Environmental Monitoring and Protection
As environmental concerns grow in importance, smart materials that adapt to changes in pollution, temperature, and humidity are gaining popularity. Smart materials-based sensors may be constructed to detect environmental changes in real time, providing novel options for pollution monitoring, water purification, and waste management. Future studies should look on ways to increase these materials' sensitivity, selectivity, and endurance across a wide range of environmental circumstances.

v. AI and Smart Material Design

The use of artificial intelligence (AI) and machine learning (ML) in smart material design is a promising trend. AI-powered models can anticipate material behavior, optimize characteristics, and speed up the development of novel materials with desired functionality. Research in this field might concentrate on developing more complex computational frameworks that enable real-time modifications in smart materials, resulting in adaptive systems for healthcare, robotics, and smart infrastructure.

11.6 Conclusions

In this comprehensive review of emerging trends and impending developments in smart materials, we have explored the transformative impact these materials are having across various domains. Smart materials, categorized by their ability to respond dynamically to external stimuli, are at the forefront of technological innovation, offering enhanced functionality and adaptability in diverse applications. The incorporation of advanced multifunctional polymer compounds, which include shape memory polymers (SMPs) and photo-responsive polymers, has revolutionized numerous industries. SMPs, with their distinctive ability to revert to a predefined shape in response to an external stimulus, are proving advantageous for medical equipment, textiles and clothing, aerospace engineering, and the field of robotics. Their portability, relatively inexpensive cost, and biocompatibility make them appropriate for a wide range of applications, including stents and adaptable clothing items, as well as detachable aircraft structures and self-healing materials. The continued development of SMPs offers even more versatility and performance, expanding their value in new and practical applications. Photochromic, photothermal, photomechanical, and photo-responsive polymers, on the other hand, offer intriguing opportunities through their responsiveness to luminescent stimuli, enabling innovations in smart coatings and sensors, where their capacity to shifts properties in response to light enhances features such as flexible color transformation in coatings and highly sensitive light detection in sensors.

Looking towards the future, the continued advancement of smart materials will be driven by the need for more efficient, adaptable, and multifunctional solutions across various sectors. Innovations in materials science, coupled with interdisciplinary approaches, will further enhance the capabilities of smart materials. For instance, the integration of carbon nanomaterials, such as graphene and carbon nanotubes, with traditional smart materials holds promise for developing high-performance materials with improved mechanical, electrical, and thermal properties. In conclusion, the field of smart materials is rapidly developing, with significant strides being made in understanding and harnessing their properties. As technology improves, we should anticipate more complex and profound applications to emerge. The continued development of intelligent materials will address significant technological and societal concerns, paving the path for breakthroughs that improve functionality, sustainability, and overall quality of existence. The combination of material science and emerging technologies is expected to fuel the subsequent era of innovation, emphasizing the significance of innovative materials in influencing the future of many industries.

References

1. Mazari, S.A., Ali, E., Abro, R., Khan, F.S.A., Ahmed, I., Ahmed, M., Nizamuddin, S., Siddiqui, T.H., Hossain, N., Mubarak, N.M., Shah, A., Nanomaterials: Applications, waste-handling, environmental toxicities, and future challenges–A review. *J. Environ. Chem. Eng.*, *9*, 2, 105028, 2021.
2. Zhang, J., Tang, W., Zhang, X., Song, Z., Tong, T., An overview of stimuli-responsive intelligent antibacterial nanomaterials. *Pharmaceutics*, 15, 8, 2113, 2023.
3. Su, M. and Song, Y., Printable smart materials and devices: strategies and applications. *Chem. Rev.*, *122*, 5, 5144–5164, 2021.
4. Lazurko, C., Jacques, E., Ahumada, M., Alarcon, E.I., Nanomaterials for its use in biomedicine: An overview, in: *Nanoengineering Materials for Biomedical Uses*, pp. 1–11, 2019.
5. de Lima, C.R., Vatanabe, S.L., Choi, A., Nakasone, P.H., Pires, R.F., Silva, E.C.N., A biomimetic piezoelectric pump: Computational and experimental characterization. *Sens. Actuators, A*, *152*, 1, 110–118, 2009.
6. Yu, X., Cheng, H., Zhang, M., Zhao, Y., Qu, L., Shi, G., Graphene-based smart materials. *Nat. Rev. Mater.*, *2*, 9, 1–13, 2017.
7. Baughman, R.H., Cui, C., Zakhidov, A.A., Iqbal, Z., Barisci, J.N., Spinks, G.M., Wallace, G.G., Mazzoldi, A., De Rossi, D., Rinzler, A.G., Jaschinski, O., Carbon nanotube actuators. *Science*, 284, 1340–1344, 1999.
8. Pelrine, R., Kornbluh, R., Pei, Q., Joseph, J., High-speed electrically actuated elastomers with strain greater than 100%. *Science*, 287, 5454, 836–839, 2000.
9. Hassani, F.A., Shi, Q., Wen, F., He, T., Haroun, A., Yang, Y., Feng, Y., Lee, C., Smart materials for smart healthcare–moving from sensors and actuators to self-sustained nanoenergy nanosystems. *Smart Mater. Med.*, *1*, 92–124, 2020.
10. Das, S., Mitra, S., Khurana, S.P., Debnath, N., Nanomaterials for biomedical applications. *Front. Life Sci.*, *7*, 3–4, 90–98, 2013.

11. Salata, O.V., Applications of nanoparticles in biology and medicine. *J. Nanobiotechnol.*, *2*, 1–6, 2004.
12. Padmanabhan, P., Kumar, A., Kumar, S., Chaudhary, R.K., Gulyás, B., Nanoparticles in practice for molecular-imaging applications: an overview. *Acta Biomater.*, 41, 1–16, 2016.
13. Genchi, G.G., Marino, A., Tapeinos, C., Ciofani, G., Smart materials meet multifunctional biomedical devices: Current and prospective implications for nanomedicine. *Front. Bioeng. Biotechnol.*, *5*, 80, 2017.
14. Cabane, E., Zhang, X., Langowska, K., Palivan, C.G., Meier, W., Stimuli-responsive polymers and their applications in nanomedicine. *Biointerphases*, *7*, 1, 9, 2012.
15. Aflori, M., Smart nanomaterials for biomedical applications—a review. *Nanomaterials*, *11*, 2, 396, 2021.
16. Jani, J.M., Leary, M., Subic, A., Gibson, M.A., A review of shape memory alloy research, applications and opportunities. *Mater. Des. (1980-2015)*, *56*, 1078–1113, 2014.
17. Lendlein, A. and Kelch, S., Shape-memory polymers. *Angew. Chem. Int. Ed.*, *41*, 12, 2034–2057, 2002.
18. Cohades, A. and Michaud, V., Shape memory alloys in fibre-reinforced polymer composites. *Adv. Ind. Eng. Polym. Res.*, *1*, 1, 66–81, 2018.
19. Huang, Y., Liang, J., Chen, Y., The application of graphene based materials for actuators. *J. Mater. Chem.*, 22, 9, 3671–3679, 2012.
20. Zhang, J., Song, L., Zhang, Z., Chen, N., Qu, L., Environmentally responsive graphene systems. *Small*, 10, 11, 2151–2164, 2014.
21. Zhao, F., Zhao, Y., Chen, N., Qu, L., Stimuli-deformable graphene materials: from nanosheet to macroscopic assembly. *Mater. Today*, 19, 3, 146–156, 2016.
22. Dayyoub, T., Maksimkin, A.V., Filippova, O.V., Tcherdyntsev, V.V., Telyshev, D.V., Shape memory polymers as smart materials: a review. *Polymers*, *14*, 17, 3511, 2022.
23. Ratna, D. and Karger-Kocsis, J., Recent advances in shape memory polymers and composites: a review. *J. Mater. Sci.*, *43*, 254–269, 2008.
24. Orozco, F., Kaveh, M., Santosa, D.S., Lima, G.M.R., Gomes, D.R., Pei, Y., Araya-Hermosilla, R., Moreno-Villoslada, I., Picchioni, F., Bose, R.K., Electroactive self-healing shape memory polymer composites based on Diels–Alder chemistry. *ACS Appl. Polym. Mater.*, *3*, 12, 6147–6156, 2021.
25. Menon, A.V., Madras, G., Bose, S., The journey of self-healing and shape memory polyurethanes from bench to translational research. *Polym. Chem.*, *10*, 32, 4370–4388, 2019.
26. Chen, Y., Chen, C., Rehman, H.U., Zheng, X., Li, H., Liu, H., Hedenqvist, M.S., Shape-memory polymeric artificial muscles: Mechanisms, applications and challenges. *Molecules*, *25*, 18, 4246, 2020.

27. Cao, L., Wang, L., Zhou, C., Hu, X., Fang, L., Ni, Y., Lu, C., Xu, Z., Surface Structures, Particles, and Fibers of Shape-Memory Polymers at Micro-/Nanoscale. *Adv. Polym. Technol., 2020*, 1, 7639724, 2020.
28. Parhi, R., Emerging applications of stimuli-responsive polymers in pharmaceutical and biomedical field. *Charact. Appl. Nanomater.*, *6*, 2, 4479, 2023.
29. Wang, L., Bisoyi, H.K., Zheng, Z., Gutierrez-Cuevas, K.G., Singh, G., Kumar, S., Bunning, T.J., Li, Q., Stimuli-directed self-organized chiral superstructures for adaptive windows enabled by mesogen-functionalized graphene. *Mater. Today*, *20*, 5, 230–237, 2017.
30. Wong, W.K., Lai, C.H.N., Cheng, W.Y., Tung, L.H., Chang, R.C.C., Leung, F.K.C., Polymer–Metal Composite Healthcare Materials: From Nano to Device Scale. *J. Compos. Sci.*, *6*, 8, 218, 2022.
31. Chadha, U., Bhardwaj, P., Agarwal, R., Rawat, P., Agarwal, R., Gupta, I., Panjwani, M., Singh, S., Ahuja, C., Selvaraj, S.K., Banavoth, M., Recent progress and growth in biosensors technology: A critical review. *J. Ind. Eng. Chem.*, *109*, 21–51, 2022.
32. Yao, X., Li, T., Wang, J., Ma, X., Tian, H., Recent progress in photoswitchable supramolecular self-assembling systems. *Adv. Opt. Mater.*, *4*, 9, 1322–1349, 2016.
33. Rasool, A., Hafeez, S., Islam, A., Khan, S.U., Fasih, M.S., Sharif, A., Khan, S.M., Hussain, N., Polymer nanocomposite films and coatings for biomedical applications, in: *Polymer Nanocomposite Films and Coatings*, pp. 729–758, Woodhead Publishing, 2024.
34. Zhao, C., Kang, J., Li, Y., Wang, Y., Tang, X., Jiang, Z., Carbon-based stimuli-responsive nanomaterials: classification and application, in: *Cyborg and Bionic Systems*, vol. *4*, 2023.
35. Hwang, H.S., Jeong, J.W., Kim, Y.A., Chang, M., Carbon nanomaterials as versatile platforms for biosensing applications. *Micromachines*, *11*, 9, 814, 2020.
36. Cha, C., Shin, S.R., Annabi, N., Dokmeci, M.R., Khademhosseini, A., Carbon-based nanomaterials: multifunctional materials for biomedical engineering. *ACS Nano*, *7*, 4, 2891–2897, 2013.
37. Xu, Q., Li, W., Ding, L., Yang, W., Xiao, H., Ong, W.J., Function-driven engineering of 1D carbon nanotubes and 0D carbon dots: mechanism, properties and applications. *Nanoscale*, *11*, 4, 1475–1504, 2019.
38. Scida, K., Stege, P.W., Haby, G., Messina, G.A., García, C.D., Recent applications of carbon-based nanomaterials in analytical chemistry: critical review. *Anal. Chim. Acta*, *691*, 1–2, 6–17, 2011.
39. Maiti, D., Tong, X., Mou, X., Yang, K., Carbon-based nanomaterials for biomedical applications: a recent study. *Front. Pharmacol.*, *9*, 1401, 2019.
40. Zhang, B.T., Zheng, X., Li, H.F., Lin, J.M., Application of carbon-based nanomaterials in sample preparation: A review. *Anal. Chim. Acta*, *784*, 1–17, 2013.

41. Jariwala, D., Sangwan, V.K., Lauhon, L.J., Marks, T.J., Hersam, M.C., Carbon nanomaterials for electronics, optoelectronics, photovoltaics, and sensing. *Chem. Soc. Rev.*, *42*, 7, 2824–2860, 2013.
42. Yang, C., Denno, M.E., Pyakurel, P., Venton, B.J., Recent trends in carbon nanomaterial-based electrochemical sensors for biomolecules: A review. *Anal. Chim. Acta*, 887, 17–37, 2015.
43. Geng, H., Peng, Y., Qu, L., Zhang, H., Wu, M., Structure design and composition engineering of carbon-based nanomaterials for lithium energy storage. *Adv. Energy Mater.*, *10*, 10, 1903030, 2020.
44. Wu, Q., Yang, L., Wang, X., Hu, Z., From carbon-based nanotubes to nanocages for advanced energy conversion and storage. *Acc. Chem. Res.*, *50*, 2, 435–444, 2017.
45. Nasrollahzadeh, M., Sajjadi, M., Iravani, S., Varma, R.S., Carbon-based sustainable nanomaterials for water treatment: state-of-art and future perspectives. *Chemosphere*, *263*, 128005, 2021.
46. Thines, R.K., Mubarak, N.M., Nizamuddin, S., Sahu, J.N., Abdullah, E.C., Ganesan, P., Application potential of carbon nanomaterials in water and wastewater treatment: a review. *J. Taiwan Inst. Chem. Eng.*, *72*, 116–133, 2017.
47. Eivazzadeh-Keihan, R., Noruzi, E.B., Chidar, E., Jafari, M., Davoodi, F., Kashtiaray, A., Gorab, M.G., Hashemi, S.M., Javanshir, S., Cohan, R.A., Maleki, A., Applications of carbon-based conductive nanomaterials in biosensors. *Chem. Eng. J.*, *442*, 136183, 2022.
48. Wen, J., Xu, Y., Li, H., Lu, A., Sun, S., Recent applications of carbon nanomaterials in fluorescence biosensing and bioimaging. *Chem. Commun.*, *51*, 57, 11346–11358, 2015.
49. Lin, J., Huang, Y., Huang, P., Graphene-based nanomaterials in bioimaging, in: *Biomedical Applications of Functionalized Nanomaterials*, pp. 247–287, 2018.
50. Eldin, M.H., Gamal, O., Mohamed, A., El-Sherbiny, I.M., Stimuli-responsive nanosystems for smart drug delivery, in: *Novel Formulations and Future Trends*, pp. 593–618, Academic Press, 2024.
51. Nabitabar, M., Shaterian, M., Danafar, H., Enhessari, M., Multi-wall carbon Nanotube surface-based functional nanoparticles for stimuli-responsive dual pharmaceutical compound delivery. *Sci. Rep.*, *14*, 1, 12073, 2024.

12

Health Empowering Materials and Its Applications in Medical Healthcare

N. Thiyagarajan[1*], V. Loganathan[2], K. Vishnulakshmi[3], Veena Raj[4] and K. Swathy[5]

[1]*Department of Electronics and Communication Engineering, KGISL Institute of Technology, Coimbatore, India*
[2]*Department of Biomedical Engineering, Sona College of Technology, Salem, India*
[3]*Department of Biomedical Engineering, DR. N.G.P. Institute of Technology, Coimbatore, India*
[4]*Faculty of Integrated Technologies, Universiti Brunei Darussalam, Bandar Seri Begawan, Brunei Darussalam*
[5]*Department of Biomedical Engineering, Dhanalakshmi Srinivasan College of Engineering, Coimbatore, India*

Abstract
The revolution in materials advances transformative healthcare practices. In recent years, the synergy between materials science and medical innovation has led to remarkable advancements by offering novel solutions to longstanding challenges in diagnosis, treatment, and patient care. This chapter summarizes the most recent developments in this area, emphasizing the creation and use of innovative materials in tissue engineering, drug delivery systems, medical devices, and diagnostic instruments. The discussion starts with the role of biomaterials in medical device design, highlighting the properties and functionalities required for successful implantation and integration within the human body. It explores the use of biocompatible polymers, metals, ceramics, and composites in fabricating orthopedic implants, cardiovascular stents, and prosthetic devices. This is to emphasize the importance of material selection in ensuring biocompatibility and long-term performance. The chapter delves into the use of scaffold materials and biodegradable polymers to promote organ transplantation and tissue regeneration. It talks about the difficulties in replicating the intricate structure and functionality of natural

**Corresponding author*: thiyaguec@gmail.com

Kolla Bhanu Prakash, S. V. Ranganayakulu and K. S. Jagannatha Rao (eds.) New Frontiers in Materials Science: Interdisciplinary Approaches to Innovation and Technologies, (241–260) © 2025 Scrivener Publishing LLC

tissues as well as the most current developments in bio-fabrication methods for the creation of functional tissue constructs, such as 3D bioprinting and microfluidic technologies. It also addresses the importance of biocompatibility, durability, and safety considerations in the design and implementation of medical materials, ensuring their efficacy and suitability for clinical applications. To facilitate early disease identification and monitoring, the functions of innovative materials in diagnostic imaging and biosensing are reviewed. Potential transformative materials shape the future of medicine, fostering innovation and empowering health globally. By illuminating the diverse roles and applications of materials in healthcare, this chapter underscores their pivotal contribution to improving patient outcomes, enhancing treatment modalities, and advancing the overall quality of healthcare delivery.

Keywords: Biomaterials, healthcare, medical device design, biocompatibility, microfluidic technologies

12.1 Introduction

A novel strategy for enhancing healthcare outcomes through the application of sophisticated materials science is represented by health-empowering materials. Personalized care, rehabilitation, diagnosis, and treatment are the prominent areas in which these materials are essential for faster patient recovery. Researchers and healthcare experts hope to improve patient outcomes, maximize treatment effectiveness, and completely transform healthcare delivery by fusing cutting-edge materials with biomedical technology [1]. Biomaterials are materials designed to work with biological systems in a medical setting. Examples of applications employing biomaterials include medication delivery systems, scaffolds for tissue engineering, and implants. For tissue repair, replacement, and regeneration, these biomaterials are used [2].

Nanomaterials are one of the most interesting choices for a biomedical application. In recent times, there have been many improvements in nanotechnology, leading to drug delivery and high-contrast imaging [3]. Biomaterials have been developed based on the functions and structure of living organisms. Self-healing and biocompatibility are the most important qualities of these materials, following the principles of natural design [4]. In the healthcare industry, smart materials are used as wearable technology such as for diagnostics, implants, and during patient care, it is used for monitoring and feedback purposes [5]. The figure depicted in Figure 12.1 shows how biomaterials can be selected for bio-implantable tissues.

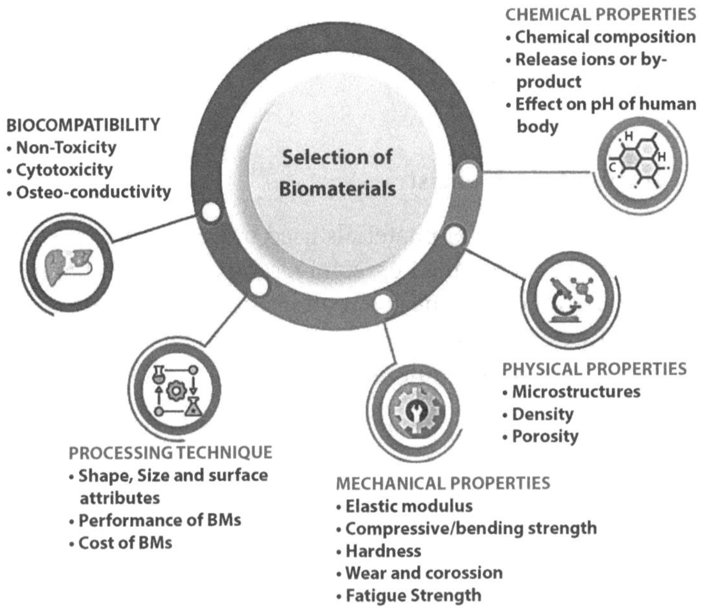

Figure 12.1 Biomaterials for orthopedic implants [7].

12.1.1 Medical Implants Made of Biomaterials

Biomaterials are the most fundamental factor for medical implants in the healthcare industry. These implants are used to repair or replace damaged tissues and also to naturally support the healing process of the body. Biomaterials have specific characteristics such as biocompatibility, adjustable degradation rates, and mechanical robustness. With the help of these biomaterial characteristics, medical implants have enhanced the patient's life span [6].

12.1.2 Implants for Orthopedics

Biomaterials are used in joint replacements, spinal fusion, and fracture repair. Some of the polymers used in orthopedic implants are titanium alloys and biodegradable polymers. These orthopedic implants enhance the patient's life span by restoring movement and encouraging bone regeneration [8].

12.1.3 Cardiovascular Implants

Biomaterials are used in cardiac implants to place pacemakers, heart valves, and stents according to patients' needs. The consequences of mild heart attacks and strokes are controlled by blood flow restoration with the help of biocompatible stents [9]. Regenerative materials are used for heart valve creation, and they can be replaced for heart failure patients to improve the patient's quality of life [10].

12.1.4 Implants in Dentistry

Titanium and zirconia are biomaterials used in dental implants to replace oral failures by a process called osseointegration. These implants coordinate with the surrounding bone tissue and support dental prostheses such as bridges, dentures, and crowns. They ensure patients' oral health, giving them the confidence to eat, smile, and speak.

12.1.5 Brain Implants

Biomaterials are used in deep brain stimulators, spinal cord stimulators, and cochlear implants. Patients who are facing difficulties in movement, chronic pain symptoms, and hearing loss can be relieved by these implants through brain activity and electrical stimulation [11]. These implants also improve the neuronal function of a patient.

12.1.6 Implants Made of Soft Tissue

Biomaterials such as biodegradable polymers and hydrogels are used in hernia repair, drug delivery, and breast reconstruction. The use of these materials in tissue engineering helps speed up the patient's recovery time and also results in better cosmetic outcomes [12].

12.2 Drug Delivery Technology and Nanotechnology

This is the most important factor in the healthcare industry. It comprises several innovative methods for delivering quality medication to patients. By using the nanoparticles, we can achieve more sustainable kinetics, and it also facilitates personalized methods. Figure 12.2 illustrates nanotechology-based drug delivery system.

12.2.1 Drug Administration Systems Based on Nanoparticles

Nanoparticles such as liposomes, dendrimers, and polymeric nanoparticles are used for drug delivery management because of their increased

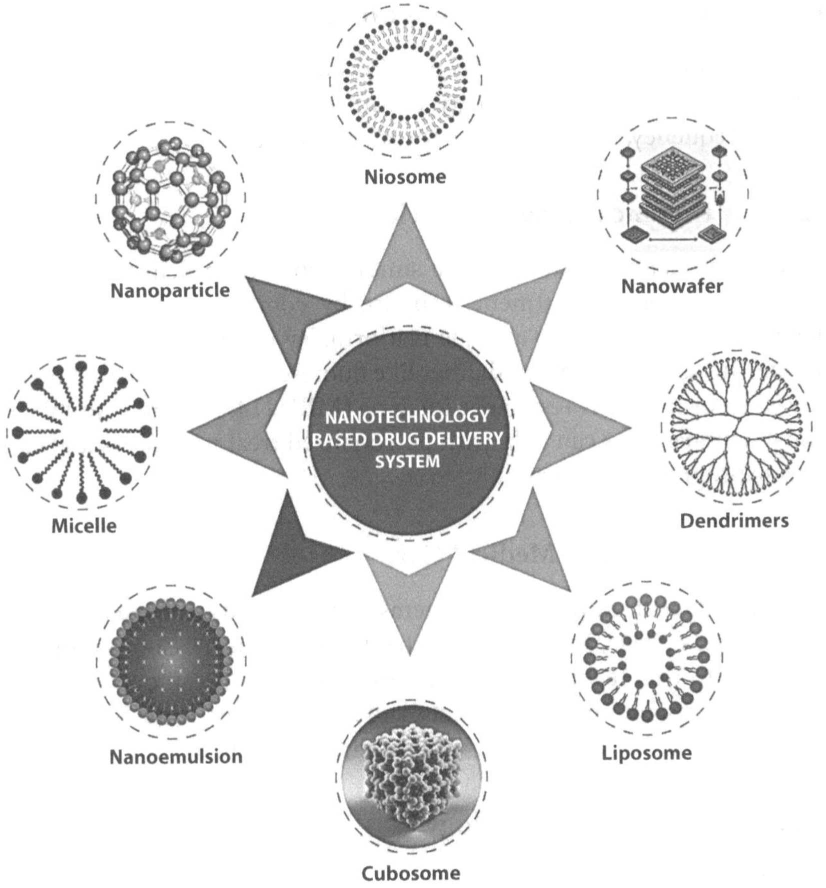

Figure 12.2 Nanotechnology-based drug delivery systems [13].

surface area-to-volume ratio and tiny size. There is a possibility to increase the therapeutic efficacy and decrease systemic toxicity by delivering drugs to targeted tissues by adding ligands to the nanoparticles.

12.2.2 Targeted Drug Delivery

Nanotechnology helps in drug delivery to sick cells while preserving healthy tissues. Drug delivery that reduces off-target effects and is also site-specific is made possible by functionalized nanoparticles. These nanoparticles have the ability to identify and bind to molecular targets that are placed on the surface of cancer cells and compromised organs.

12.2.3 Managed Release Frameworks

By maintaining therapeutic concentrations within the therapeutic window for optimal efficacy, this strategy optimizes drug bioavailability, lowers dosage frequency, and increases patient compliance.

12.2.4 Theranostic Nanoparticles

Theranostic nanoparticles provide simultaneous drug delivery and real-time monitoring of treatment response by combining therapeutic and diagnostic functions onto a single platform. By including imaging agents for non-invasive imaging modalities like fluorescence, computed tomography (CT), or magnetic resonance imaging (MRI) [14, 15], these multifunctional nanoparticles can help with personalized treatment plans and early disease progression detection.

12.2.5 Customized Medical Care

Customized medication delivery systems that are based on a patient's genetic profile, disease stage, and response to treatment can be developed more easily thanks to nanotechnology. Healthcare professionals can optimize treatment regimens, minimize side effects, and improve therapeutic outcomes by designing drug delivery systems with patient-specific information in mind.

12.2.6 Novel Nanotechnology

Advances in nanotechnology could revolutionize drug delivery in medicine with innovations such as stimuli-responsive nanoparticles, exosome-based drug delivery devices, and DNA origami nanorobots [16]. These cutting-edge technologies provide new approaches to improving medication targeting, overcoming biological barriers, and having exact control over the kinetics of drug release.

12.3 Regenerative Medicine and Tissue Engineering

Through the application of biomaterials, stem cells, growth factors, and other physiologically active compounds, these interdisciplinary fields seek to maintain, improve, or restore tissue function. Figure 12.3 depicts the technologies employed in regenerative medicine.

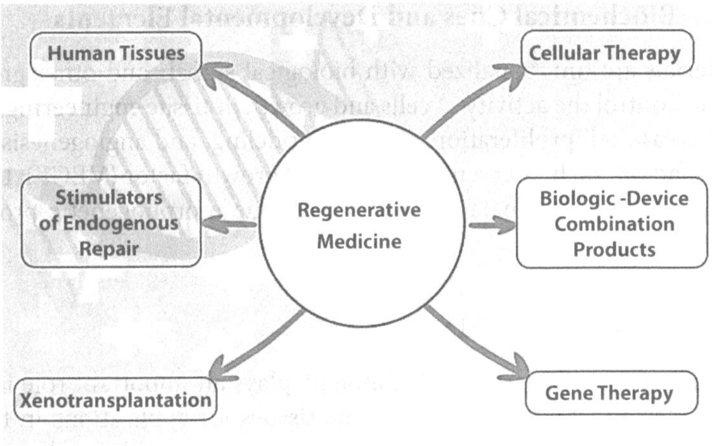

Figure 12.3 Major technologies used in regenerative medicine [17].

Tissue engineering and regenerative medicine allow patients to heal wounds, replace damaged tissue, and regain physiological function by utilizing cutting-edge methods in materials science and biomedical engineering.

12.3.1 Biomaterial Ensembles

Structurally supporting cell adhesion, proliferation, and differentiation, biomaterial scaffolds also function as templates for tissue regeneration [18]. Extracellular matrix (ECM) that has been decellularized from biological tissues, natural polymers like collagen and hyaluronic acid, or synthetic polymers are examples of biocompatible and biodegradable materials that are commonly used to make these scaffolds [17]. Biomaterial scaffolds promote the integration of surrounding host tissues with tissue regeneration by imitating the native tissue microenvironment.

12.3.2 Using Stem Cells in Therapy

Stem cells are special because they can develop into numerous kinds of cells and repair damaged tissues. Stem cell therapy has a high impact on tissue engineering for treating wounds, such as musculoskeletal injuries, cardiovascular illnesses, and neurological diseases [19]. Tissue regeneration stem cell therapy is enhanced by biomaterial scaffolds.

12.3.3 Biochemical Cues and Developmental Elements

Biomaterials are functionalized with biological signals and other growth factors to control the activity of cells and growth in tissue engineering. This helps increase cell proliferation, tissue remodeling, and angiogenesis, and includes factors such as vascular endothelial growth factor (VEGF), transforming growth factor-beta (TGF-β), and bone morphogenetic proteins (BMPs).

12.3.4 Bioprinting in 3D

Three-dimensional bioprinting technology plays an important role in tissue engineering which aims to fabricate tissues for applications in tissue engineering and the testing or screening of drugs. Additionally, tissue regeneration and reconstruction enable the possibility to repair or replace damaged tissues and organs [20, 21].

12.3.5 Regeneration of Organs

The main aim of tissue engineering is to repair damaged tissues and organs with the help of the body's natural ability for regeneration. Potential remedies for organ failure and tissue loss include tissue engineering, organoid culture, and cell transplantation. In artificial tissues and organ constructs, biomaterials are essential for providing structural support, encouraging cell integration, and aiding vascularization [22].

12.3.6 Regenerative Therapies and Clinical Translation

Innovative treatments for a variety of medical illnesses, such as chronic wounds, osteoarthritis, spinal cord injuries, and heart ailments, have been developed as a result of advancements in tissue engineering and adaptive medicine. These treatments have the potential to address unmet clinical needs in various medical sectors, reduce healthcare costs, and improve patient outcomes.

12.4 Types of Nanoparticles

There are numerous varieties of nanoparticles accessible for use in medical applications. A few types of particles are discussed below:

12.4.1 Micelles

Micelles are generally amphiphilic surfactant molecules that can be formed by combining lipids and other amphiphilic molecules. Hydrophobic therapeutic drugs can be encapsulated into micellar spherical vesicles, which spontaneously aggregate and self-assemble into a hydrophilic outer monolayer and a hydrophobic core in aqueous environments. Due to the special qualities of micelles, hydrophobic medications can be rendered more soluble, increasing their bioavailability. Micelles typically have sizes between 10 and 100 nm. Micelles can be utilized for therapeutic, imaging, contrast, and drug delivery agents, among other purposes.

12.4.2 Liposomes

Lipid bilayers form the basis of spherical vesicles called liposomes, which have particle sizes varying from 30 nanometers to several microns. Therapeutic substances that are hydrophilic in the aqueous phase and hydrophobic in the liposomal membrane layer can be incorporated into liposomes. Because liposomes are so adaptable, macromolecular medications such as crystalline metals and nucleic acids can be incorporated into them by changing the surface properties of the liposomes using polymers, antibodies, and/or proteins.

12.4.3 Dendrimers

Macrophages known as dendrimers have external operational groups and split off repetitive units that radiate out from a central core. These operational groups can change an anionic, neutral, or cationic terminal's overall structure, as well as its chemical and physical properties. Dendrimers are highly bioavailable and biodegradable because therapeutic substances can be incorporated into their inner spaces or affixed to their surface groups. Upon absorption of medicinal medicines, dendrimer conjugates containing saccharides or peptides have demonstrated better antibacterial, antiprion, and antiviral capabilities, along with enhanced solubility and stability. Polyamidoamine dendrimer-DNA complexes also referred to as dendriplexes, have been studied as potential vectors for drug delivery and show promise in promoting targeted drug delivery, subsequent gene expression, and enhanced medication efficacy. Dendrimers show great potential.

12.4.4 Carbon Nanotubes

Carbon nanotubes are cylindrical molecules composed of sheets of carbon atoms gathered into a single layer (graphene). They may consist of many concentrically connected nanotubes or have one or more walls. Carbon nanotubes can attain remarkably large loading capacities as medication carriers because of their large exterior surface area. Additionally, carbon nanotubes have unique optical, mechanical, and electrical properties that make them appealing as biological sensors and imaging contrast agents.

12.4.5 Metallic Nanoparticles

Gold and iron oxide are examples of metallic nanoparticles. Hydrophilic polymers like poly ethylene glycol (PEG) or dextran are combined with a magnetic core measuring 4–5 nm to form iron oxide nanoparticles. On the other hand, surface moieties can be introduced in the form of a monolayer as ligands for active targeting to functionalize gold nanoparticles, which consist of a core of gold atoms surrounded by groups that react badly on the surface. Metallic nanoparticles have been used in optical biosensors, medicinal drugs, the food industry, and coating.

12.4.6 Quantum Dots

Quantum dots (QDs) are fluorescent semiconductor particles with sizes typically varying from 1 to 100 nm that confine electrons in three dimensions. They have shown promise for usage in a number of biomedical applications, including biomedicine and environment, catalysis, photovoltaic, cellular imaging, and drug delivery. The structure of quantum dots is shell-core and composed of elements from groups II–VI such as CdSe, CdS, or CdTe, groups III–V such as InP or InAs, or groups IV–VI such as PbSe. Quantum dots are unique in their optical characteristics and have been used in medical imaging due to their small size, excellent brightness, and stability.

12.5 Application of Organic Nanomaterials

12.5.1 Various Applications of Nanoliposomes

A liposome is a naturally occurring empty balloon composed of phospholipid and cholesterol bilayer. It can effectively preserve and release the loaded pharmaceuticals due to its high biocompatibility and structural similarity to the cell membrane. Its particle size, which varies from 20 to 200 nm, is controllable. Smaller drugs are easier for human lytic enzymes to digest and break down, which increases their absorption by the body. Furthermore, the distribution of medications can be targeted by the liposomes used as drug carriers, increasing efficacy and decreasing harmful and side effects. This presents a significant opportunity for the development of anticancer nanodrugs.

One type of anticancer material that was taken out of Sunglo is C18H17NO6. However, it has low solubility in water, which affects its effectiveness. Lan *et al.* [23] combined C18H17NO6 with cholesterol and soybean phospholipids to create nanoliposomes. The research conducted on tumor transplantation in naked mice demonstrated that as the duration of medication administration increased, the naked mice's relative tumor proliferation rate (RPR) decreased. Furthermore, a strong positive association was found between them. The greatest inhibition rate of cell proliferation was below 99.95% when 4 μg/ml of drug-loaded nanoliposomes were present, offering guidance for future drug research and development.

12.5.2 Graphene's Usage in the Medical Field

Under certain oxidation circumstances, graphene nanoparticles can produce graphene oxide. The material's edges and surface are abundant in groups that include oxygen, including epoxy, hydroxyl, and carbonyl groups. Owing to these properties, it is possible to create functional graphene oxide nanomaterials by altering additional surface-active groups. Additionally, graphene oxide material is biocompatible and has strong water solubility. In the field of biomedicine, it has been frequently utilized as a medication carrier.

12.5.3 Application of Metal-Organic Frame

A type of hybrid material called a metal-organic framework (MOF) is put together by organic ligands that are coupled to metal ions or clusters. Some of its characteristics include a large specific surface area, porous and adjustable pore size, good biocompatibility, customizable composition, modifiable surface, etc., which together possess a great deal of potential for developing drug delivery carriers in the field of cancer therapy.

The interaction between pharmacological molecules and MOFs can be improved by the particular open metal sites and MOFs' organic functional groups, resulting in regulated drug release and increased delivery efficiency. Due to the variations in pore shape and drug activity in the simulated human environment, MIL-53/100/101 has a release time of 20, 3, and 6 days, respectively. MIL-101 has four times the drug content of MIL-100. It is important to remember that MIL-53 is a 100% drug-releasing flexible mesoporous material.

Studies have also demonstrated the significant potential for the development of zeolitic imidazolate frameworks (ZIFS) in small molecule therapeutic delivery. One of these, ZIF-8, is a metal framework with a large, controlled aperture that is made of zinc and 2-methylimidazole. The Zn-DOX-ZIF-8 nanoparticle, which was produced by Zheng *et al.* [24], has a stronger curative impact on breast cancer than the DOX medication alone. This material has good biocompatibility, good dispersion and stability, and good pH responsiveness.

12.6 Embedded and Wearable Sensors

Embedded sensors play a vital role in empowering patient health in the medical field. These sensors are used for real-time monitoring of environmental conditions, physiological indicators, and illness biomarkers. They are also useful for providing personalized treatment for the patient.

12.6.1 Monitoring Patients Remotely

The wearable sensors remotely monitor the patient's health factors such as blood pressure, respiration rate, temperature, and heart rate. These sensors provide continuous health status monitoring, which helps patients improve recovery from surgery and manage chronic illnesses.

12.6.2 Health and Wellness Tracking

These sensors are used for tracking patients' general health factors like sleep patterns, stress levels, state of well-being, and exercise levels [25]. They also promote the prevention of disease, encourage healthy lifestyles, and provide real-time feedback [26]. These sensors are also used in rehabilitation and physical settings to track muscle activation, biomechanical parameters, and movement patterns. They also provide feedback for both patients and general therapists after surgery [27].

12.7 Regulatory and Ethical Considerations in Health-Empowering Materials

The application of these materials in the healthcare industry should have ethical and regulatory factors to safeguard patients.

12.7.1 Regulatory Approval Procedures

The health-improving materials are governed by international regulatory bodies such as the European Medicines Agency (EMA). Food and Drug Administration (FDA) in the post-market surveillance, preclinical research, clinical trials, and also for manufacturing standards. To assess the health-improving products for regulatory bodies, clinical usage, and efficacy data. Additionally, to enhance toxicity, immunogenicity, and biocompatibility, these factors carry out thorough preclinical research.

12.7.2 Moral Issues

Concerns regarding data security and privacy are raised by the fact that health empowerment materials frequently entail the gathering, storing, and analysis of sensitive patient data. Strong data protection mechanisms must be put in place by developers to protect patient privacy, stop illegal access to or exploitation of data, and adhere to pertinent privacy laws, such as the Health Insurance Portability and Accountability Act (HIPAA) in the US [28].

12.8 Developing Technologies and Future Paths

Medical and healthcare could be revolutionized by emerging technologies and new directions in health-promoting materials.

12.8.1 The Fields of Nanotechnology and Nanomedicine

Healthcare will continue to benefit greatly from nanotechnology since it allows for precise drug delivery, molecular imaging, and nanoscale diagnostic applications. The creation of multifunctional nanoparticles for theranostic applications and the incorporation of nanomaterials into wearable sensors and implantable devices for real-time health monitoring are potential future developments.

12.8.2 Cutting-Edge Biomaterials

The creation of new materials with improved characteristics, such as increased biocompatibility, adjustable degradation rates, and stimuli-responsive behavior, will result from ongoing developments in the field of biomaterials research. The performance of medical devices, drug delivery, and tissue regeneration will all be more precisely controlled thanks to these advancements.

12.8.3 Regenerative Medicine and Stem Cell Treatments

Research on improving stem cell sources, delivery systems, and differentiation techniques for diverse tissue regeneration applications will keep stem cell-based therapeutics evolving. The therapeutic potential of regenerative medicine approaches will be increased by developments in stem cell engineering, genetic editing, and tissue reprogramming [29].

12.8.4 Healthcare with Precision and Personalized Medicine

By combining health-empowering materials with personalized medical procedures, treatments can be customized depending on the unique characteristics of each patient, such as their genetic profile, illness subtype, and response to treatment. Implantable devices, diagnostic tools, and personalized drug delivery systems that maximize therapeutic outcomes and minimize side effects will all be made possible by biomaterials and nanotechnologies [30, 31].

12.8.5 Artificial Intelligence and Machine Learning

Data-driven decision-making, predictive analytics, and customized treatment plans will all be made easier by the use of artificial intelligence (AI) and machine learning algorithms in the healthcare industry. To find trends, improve treatment plans, and spur innovation in materials that empower people's health, AI-powered platforms will examine enormous datasets of patient data, biomaterial characteristics, and clinical outcomes.

12.9 Clinical Applications and Case Studies

The practical effects of health-promoting materials in medicine and healthcare are illustrated through case studies and clinical applications.

These examples demonstrate how cutting-edge materials and technology are being adapted into clinical practice to improve patient outcomes, increase treatment efficacy, and answer unmet medical needs across a variety of healthcare areas.

12.9.1 Implanted Health Care Equipment

To sustain cardiac function while awaiting a heart transplant, a patient suffering from severe heart failure is fitted with a left ventricular assist device (LVAD) composed of biocompatible materials [32]. The LVAD prolongs survival and enhances the patient's quality of life until a suitable donor heart is found.

12.9.2 Methods of Drug Delivery

Chemotherapy is administered to a cancer patient using drug delivery systems based on nanoparticles that are intended to specifically target tumor cells while avoiding healthy tissues. Tumor regression and increased survival rates result from the focused medication distribution, which reduces systemic toxicity and boosts chemotherapy's effectiveness. Additionally, an implantable drug-eluting device is used to administer biologic therapy to a patient suffering from rheumatoid arthritis. By releasing anti-inflammatory medications locally at the location of joint inflammation, the device lessens discomfort and inflammation while enhancing joint mobility and function [33].

12.9.3 Bioengineering and Regenerative Health Sciences

Bioengineered skin substitutes, composed of biomaterial scaffolds seeded with patient-derived cells, are used in skin regeneration therapy for patients with severe burns. These tailored skin transplants decrease scarring, promote wound healing, and restore skin function, enhancing the patient's quality of life. Bioresorbable implants composed of biocompatible polymers are used in spinal fusion surgery for a patient with degenerative disc degeneration. Implant removal surgery is not necessary, and surgical complications are decreased since the implants offer structural support as the bone heals and gradually deteriorates over time [34].

12.9.4 Devices for Wearable Health Monitoring

Wearable technology with biosensors allows patients with long-term conditions like diabetes or hypertension to track vital signs, blood sugar, and medication compliance in real-time. Continuous monitoring makes it possible to identify health problems early, facilitates prompt action easier, and gives individuals the control to take an active role in their own healthcare. An electrocardiogram (ECG) [35] sensor on a smartwatch is worn by an older patient with heart disease to track anomalies in their cardiac rhythm. When the gadget identifies an irregular heartbeat that could be an indication of atrial fibrillation, it alerts the user, urging them to seek treatment right away to avoid a stroke or other cardiovascular problems [36].

12.10 Conclusion

The revolution in materials advances transformative healthcare practices. In recent years, the synergy between materials science and medical innovation has led to remarkable advancements, offering novel solutions to longstanding challenges in diagnosis, treatment, and patient care. It is summarized that the most recent developments in this area enhance the creation and use of innovative materials in medical devices, regenerative medicine, and diagnostic instruments. The role of implantable biomaterials in medical device design highlights the properties and functionalities required for successful implantation and integration within the human body. This chapter also delivers the use of scaffold nanomaterials and biodegradable polymers to promote artificial organs and tissue engineering. It also addresses biocompatibility, durability, and safety considerations in the design and implementation of medical materials. It also discussed the function of innovative materials in diagnostic imaging and biosensing for the identification and monitoring of early-stage diseases. These transformative materials have the potential to shape the future of medicine, fostering innovation and empowering health globally.

References

1. Bohr, A. and Memarzadeh, K., The rise of artificial intelligence in healthcare applications, in: *Artificial Intelligence in Healthcare*, pp. 25–60, Academic Press, 2020, 10.1016/B978-0-12-818438-7.00002-2.
2. Junaid, S.B., Imam, A.A., Balogun, A.O., De Silva, L.C., *et al.*, Recent Advancements in Emerging Technologies for Healthcare Management Systems: A Survey. *Healthcare (Basel)*, 10, 10, 1940, 2022, doi: 10.3390/healthcare10101940.

3. Haleem, A., Javaid, M., Singh, R.P., Suman, R., Medical 4.0 technologies for healthcare: Features, capabilities, and applications. *Internet Things Cyber-Phys. Syst.*, 2, 100027, 2022, https://doi.org/10.1016/j.iotcps.2022.04.001.
4. Musah, S. and Arzaghi, H., Unleashing the power of biomaterials to enhance organoid differentiation and function. *Nat. Methods*, 21, 1575–1577, 2024, https://doi.org/10.1038/s41592-024-02393-5.
5. Lokwani, R., Josyula, A., Ngo, T.B., *et al.*, Pro-regenerative biomaterials recruit immunoregulatory dendritic cells after traumatic injury. *Nat. Mater.*, 23, 147–157, 2024, https://doi.org/10.1038/s41563-023-01689-9.
6. Nguyen, M., Karkanitsa, M., Christman, K.L., Design and translation of injectable biomaterials. *Nat. Rev. Bioeng.*, 2, 810–828, 2024, https://doi.org/10.1038/s44222-024-00213-1.
7. Jones, L.C., Topoleski, L.D.T., Tsao, A.K., Biomaterials in orthopaedic implants, in: *Mechanical Testing of Orthopaedic Implants*, pp. 17–32, Elsevier, Woodhead Publishing, 2017, https://doi.org/10.1016/B978-0-08-100286-5.00002-0.
8. Bonferoni, M.C., Caramella, C., Catenacci, L., Conti, B., *et al.*, Biomaterials for Soft Tissue Repair and Regeneration: A Focus on Italian Research in the Field. *Pharmaceutics*, 13, 9, 1341, 2021, doi: 10.3390/pharmaceutics13091341.
9. Shiravand, A., Richter, K., Willmann, P., *et al.*, Fabrication, characterization and numerical validation of a novel thin-wall hydrogel vessel model for cardiovascular research based on a patient-specific stenotic carotid artery bifurcation. *Sci. Rep.*, 14, 16301, 2024, https://doi.org/10.1038/s41598-024-66777-5.
10. Al-Shalawi, F.D., Mohamed Ariff, A.H., Jung, D.-W., Mohd Ariffin, M.K.A., *et al.*, Biomaterials as Implants in the Orthopedic Field for Regenerative Medicine: Metal versus Synthetic Polymers. *Polymers*, 15, 12, 2601, 2023, https://doi.org/10.3390/polym15122601.
11. Wood, H., Stem cell implants show promise in chronic traumatic brain injury. *Nat. Rev. Neurol.*, 17, 64, 2021, https://doi.org/10.1038/s41582-021-00459-y.
12. He, N., Wang, X., Shi, L., *et al.*, Photoinhibiting via simultaneous photoabsorption and free-radical reaction for high-fidelity light-based bioprinting. *Nat. Commun.*, 14, 3063, 2023, https://doi.org/10.1038/s41467-023-38838-2.
13. Todros, S., Todesco, M., Bagno, A., Biomaterials and Their Biomedical Applications: From Replacement to Regeneration. *Processes*, 9, 11, 1949, 2021, https://doi.org/10.3390/pr9111949.
14. Patra, J.K., Das, G., Fraceto, L.F., *et al.*, Nano based drug delivery systems: recent developments and future prospects. *J. Nanobiotechnol.*, 16, 71, 2018, https://doi.org/10.1186/s12951-018-0392-8.
15. Rizvi, S.A.A. and Saleh, A.M., Applications of nanoparticle systems in drug delivery technology. *Saudi Pharm. J.*, 26, 1, 64–70, 2018 Jan, doi: 10.1016/j.jsps.2017.10.012.
16. Yusuf, A., Almotairy, A.R.Z., Henidi, H., Alshehri, O.Y., Aldughaim, M.S., Nanoparticles as Drug Delivery Systems: A Review of the Implication of

Nanoparticles' Physicochemical Properties on Responses in Biological Systems. *Polymers (Basel)*, 15, 7, 1596, 2023 Mar 23, doi: 10.3390/polym15071596.
17. Shukla, P., Sharma, S., Rao, P., Nanoparticulate drug delivery systems: A revolution in design and development of drugs. *J. Drug Delivery Ther.*, 11, 5-S, 188–193, 2021, https://doi.org/10.22270/jddt.v11i5-S.5023.
18. Ganey, T. and Temple, H.T., Introduction to Regenerative Medicine, in: *Regenerative Medicine*, C.W. Hunter, T.T. Davis, M.J. DePalma (Eds.), Springer, Cham, 2023, https://doi.org/10.1007/978-3-030-75517-1_1.
19. Zhang, Z., Gupte, M.J., Ma, P.X., Biomaterials and stem cells for tissue engineering. *Expert Opin. Biol. Ther.*, 13, 4, 527–40, 2013, doi: 10.1517/14712598.2013.756468.
20. Watt, F.M. and Driskell, R.R., The therapeutic potential of stem cells. *Philos Trans. R Soc. Lond. B Biol. Sci.*, 365, 1537, 155–63, 2010, doi: 10.1098/rstb.2009.0149.
21. Jain, A. and Bansal, R., Applications of regenerative medicine in organ transplantation. *J. Pharm. Bioallied. Sci.*, 7, 3, 188–94, 2015, doi: 10.4103/0975-7406.160013.
22. Mohr-Haralampieva, D., Kappenthuler, S., Droege, M., Treating stress urinary incontinence by tissue engineering. *Nat. Rev. Bioeng.*, 2, 987–989, 2024, https://doi.org/10.1038/s44222-024-00246-6.
23. Lan, J., Overview of application of nanomaterials in medical domain. *Contrast Med. Mole. Imaging*, 2022, 5, 1–5, 2022, https://doi.org/10.1155/2022/3507383.
24. Zheng, C.C., Wang, Y., Phua, S.Z.F., *et al.*, ZnO-DOX@ZIF-8 core-shell nanomaterials for pH-responsive drug delivery. *ACS Biomater. Sci. Eng.*, 3, 10, 2223–2229, 2017. doi: 10.1021/acsbiomaterials.7b00435
25. Javaid, M., Haleem, A., Singh, R.P., Suman, R., 3D printing applications for healthcare research and development. *Global Health J.*, 6, 4, 217–226, 2022, https://doi.org/10.1016/j.glohj.2022.11.001.
26. Joudeh, N. and Linke, D., Nanoparticle classification, physicochemical properties, characterization, and applications: a comprehensive review for biologists. *J. Nanobiotechnol.*, 20, 262, 2022, https://doi.org/10.1186/s12951-022-01477-8.
27. Harish, V., Tewari, D., Gaur, M., *et al.*, Review on Nanoparticles and Nanostructured Materials: Bioimaging, Biosensing, Drug Delivery, Tissue Engineering, Antimicrobial, and Agro-Food Applications. *Nanomaterial (Basel)*, 12, 3, 457, 2022, doi: 10.3390/nano12030457.
28. Taléns-Visconti, R., Díez-Sales, O., de Julián-Ortiz, J.V., Nácher, A., Nanoliposomes in Cancer Therapy: Marketed Products and Current Clinical Trials. *Int. J. Mol. Sci.*, 23, 8, 4249, 2022, doi: 10.3390/ijms23084249.
29. Bhatia, S., Nanoparticles Types, Classification, Characterization, Fabrication Methods and Drug Delivery Applications, in: *Natural Polymer Drug Delivery Systems*, Springer, Cham, 2016, https://doi.org/10.1007/978-3-319-41129-3_2.

30. Syama, S. and Mohanan, P.V., Comprehensive Application of Graphene: Emphasis on Biomedical Concerns. *Nano Micro Lett.*, 6, 11, 2019, https://doi.org/10.1007/s40820-019-0237-5.
31. Hovan George, A., Shahul, A., Shaji, G., *A. Wearable sensors: A new way to track health and wellness*, vol. 1, pp. 15–34, 2023, https://doi.org/10.5281/zenodo.8260879.
32. Sharma, A., Badea, M., Tiwari, S., Marty, J.L., Wearable Biosensors: An Alternative and Practical Approach in Healthcare and Disease Monitoring. *Molecules*, 26, 3, 748, 2021, doi: 10.3390/molecules26030748.
33. Majumder, S., Mondal, T., Deen, M.J., Wearable Sensors for Remote Health Monitoring. *Sensors (Basel)*, 1217, 1, 130, 2017, doi: 10.3390/s17010130.
34. Tulchinsky, T.H., Ethical Issues in Public Health, in: *Case Studies in Public Health*, pp. 277–316, 2018, doi: 10.1016/B978-0-12-804571-8.00027-5.
35. Malik, S., Muhammad, K., Waheed, Y., Emerging Applications of Nanotechnology in Healthcare and Medicine. *Molecules, 28*, 18, 6624, 2023, doi: 10.3390/molecules28186624.
36. Anjum, S., Ishaque, S., Fatima, H., Farooq, W., *et al.*, Emerging Applications of Nanotechnology in Healthcare Systems: Grand Challenges and Perspectives. *Pharmaceuticals (Basel)*, 14, 8, 707, 2021, doi: 10.3390/ph14080707.

Index

Also of Interest

Check out these published and forthcoming titles in the "Next-Generation Computing and Communication Engineering" series from Scrivener Publishing

Digital Twins and Cybersecurity
Safeguarding the Future of Connected Systems
Edited by Palanichamy Naveen, R. Maheswar and U.S. Ragupathy
Published 2025. ISBN 978-1-394-27247-1

Computational Intelligent Techniques in Mechatronics
Edited by Kolla Bhanu Prakash, P. Satish Kumar, Ivan C. K. Tam, Wai Lok Woo and Vishal Jain
Published 2024. ISBN 978-1-394-17464-5

Machine Learning for Industrial Applications
By Kolla Bhanu Prakash
Published 2024. ISBN 978-1-394-26896-2

Machine Learning Paradigm for Internet of Things Applications
Edited by Shalli Rani, R. Maheswar, G. R. Kanagachidambaresan, Sachin Ahuja and Deepali Gupta
Published 2022. ISBN 978-1-119-76047-4

Data Science Handbook
A Practical Approach
By Kolla Bhanu Prakash
Published 2022. ISBN 978-1-119-85732-7

Digitization of Healthcare Data Using Blockchain
Edited by T. Poongodi, D. Sumathi, B. Balamurugan and K. S. Savita
Published 2022. ISBN 978-1-119-79185-0

Medical Imaging and Health Informatics
Edited by Tushar H. Jaware, K. Sarat Kumar, Svetlin Antonov, and Ravindra D. Badgujar
Published 2022. ISBN 978-1-119-81913-4

Advanced Analytics and Deep Learning Models
Edited by Archana Mire, Shaveta Malik and Amit Kumar Tyagi
Published 2022. ISBN 978-1-119-79175-1

Integration of Renewable Energy Sources with Smart Grids
Edited by A. Mahaboob Subahani, M. Kathiresh and G. R. Kanagachidambaresan
Published 2021. ISBN 978-1-119-75042-0

Cognitive Engineering for Next Generation Computing
A Practical Analytical Approach
Edited by Kolla Bhanu Prakash, G.R. Kanagachidambaresan, V. Srikanth and E. Vamsidhar
Published 2021. ISBN 978-1-119-71108-7

Role of Edge Analytics on Sustainable Smart City Development
Edited by G.R. Kanagachidambaresan
Published 2020. ISBN 9781119681281

www.scrivenerpublishing.com

Printed in the USA/Agawam, MA
September 16, 2026

910661.071